☑ Y0-BQH-704

THE BOOK OF

FAX

An Impartial Guide to
Buying & Using
Facsimile Machines

WITHDRAWN

Daniel Fishman
Elliot King

Ventana Press
Chapel Hill, NC

THE BOOK OF FAX
An Impartial Guide to Buying and Using Facsimile Machines

Copyright © 1988 Daniel Fishman and Elliot King

All rights reserved. This book may not be duplicated in any way without the expressed written consent of the publisher, except in the form of brief excerpts or quotations for the purposes of review. The information contained herein may not be duplicated in other books, databases or any other medium without the written consent of the publisher or author. Making copies of this book, or any portion for any purpose other than your own, is a violation of United States copyright laws.

Library of Congress Cataloging-in-Publication Data

Fishman, Daniel
    The book of FAX.

    Includes index.
    1. Facsimile transmission—Equipment and supplies—
Purchasing. I. King, Elliot. II. Title.
TK6710.F57 1988   621.38'0414   88-50307
ISBN 0-940087-22-7

Cover design by Holly Russell, Durham, NC

Cover photography by Henderson-Muir Photography, Raleigh, NC

Illustrations by Bob Murray, Durham, NC

Desktop Publishing by Jim Barnett, Primary Graphics, Chapel Hill, NC

Editing by David Klein, Chapel Hill, NC

Copyediting by Clare Rosen, Chapel Hill, NC

First Edition, First Printing
Printed in the United States of America

Ventana Press, Inc.
P.O. Box 2468
Chapel Hill, NC 27515
919/942-0220

Limits of Liability and Disclaimer of Warranty
The authors and publisher of this book have used their best efforts in preparation of this book. These efforts include the development, research and testing of the theories to determine their effectiveness. The author and publisher make no warranty of any kind, expressed or implied, with regard to the instructions and suggestions contained in this book.

The author and publisher shall not be liable in the event of incidental or consequential damages in connection with, or arising out of, the furnishing, performance or use of instructions, and/or claims of productivity gains.

Acknowledgments

The authors wish to thank Jan Ozer of Asher Technologies for his assistance, Jeffrey Charlston of Charlston, Revich & Williams for his research into the legal aspects of fax, Mark Chodos for his many hours of research, and Bill Gladstone for his help in initiating the project. We also wish to express our thanks to Anita and Sharon for their help and support during the writing of this book.

The publisher wishes to express appreciation to the following individuals who assisted in the production of this book:

Jim Barnett
Chip Henderson
Steve Muir
Bob Murray
Karen Wysocki
Frank Yarborough

About the Authors

Daniel Fishman is founder of Micro/Research, a computer and technology consulting firm in Monrovia, California, that specializes in integrating new technology to small and medium-sized businesses.

Daniel Fishman
Micro/Research
730 East Cypress Ave.
Monrovia, CA 91016
818/359-3503

Elliot King teaches communications at the University of California, San Diego, and has published more than 150 articles in computing magazines, including several on fax technology.

Elliot King
199 N. Camino Real
Suite F 272
Encinitas, CA 92024
619/436-4174

Trademark Acknowledgments

Ventana Press has made every effort to supply trademark information about the company name, product and services mentioned herein. Trademarks listed below were derived from various sources. The publisher cannot attest to the accuracy of this trademark information.

All products are trademarks or registered trademarks of their respective manufacturers.

AutoCAD, AUTODESK, INC.
Dr. Halo, MEDIA CYBERNETICS
Hercules Ramfont, HERCULES COMPUTERS
Hewlett-Packard Graphics Language, HEWLETT-PACKARD
Hewlett-Packard DeskJet, HEWLETT-PACKARD
IBM PC, INTERNATIONAL BUSINESS MACHINES
IBM, INTERNATIONAL BUSINESS MACHINES
Macintosh, APPLE COMPUTER, INC.
Magnavox Video Writer, MAGNAVOX INC.
MCI, MCI TELECOMMUNICATIONS CORPORATION
PC Paintbrush, Z-SOFT, INC.
PostScript, ADOBE SYSTEMS, INC.

Contents

Introduction

THE INFORMATION MANAGEMENT "BILL OF RIGHTS"

In the early 19th century, Baron de Rothschild launched one of the great fortunes of Europe when he became the first Londoner to learn that Napoleon had been defeated at Waterloo. Armed with that important knowledge, he invested in the stock market and reaped great rewards.

> **Getting the *right* information at the *right* time can offer a dramatic strategic advantage.**

Six months earlier, Andrew Jackson fought the Battle of New Orleans even though a peace treaty with the British had been signed. Because he hadn't received word from Washington about the treaty, the war dragged on needlessly.

Although the above examples are extreme, getting the *right* information at the *right* time to the *right* place in the *right* form can offer a dramatic strategic advantage. On the other hand, being uninformed can be a recipe for trouble.

FAX: A NEW CHANNEL OF COMMUNICATION

We have long since outgrown the messenger service Baron de Rothschild used to such advantage. In many ways we've also outgrown telex, telephones and overnight mail delivery. Those who can effectively establish and manage a complete communications arsenal—telecommunications, computers and facsimile—will enhance their chances to emerge as tomorrow's winners. To paraphrase Vince Lombardi, these days information isn't everything; it's the only thing.

WHAT IS FAX?

Contemporary facsimile equipment is an innovative technology that already plays a key role in the information and communication revolution. In brief, fax (short for *facsimile*) equipment allows you to duplicate almost any type of text or graphic information and transmit it over telephone lines to another fax machine.

> **Unlike computers, which transmit only files, fax transmits any graphic image.**

To transmit information, fax machines scan documents and convert the information into an electronic format. That code is then sent via regular telephone lines to the receiving machine. There the information is decoded and printed, precisely reproducing the original document. At that point, the recipient has a duplicate of the original, while the sender keeps the original copy.

Unlike telephones, which usually transmit only sounds, fax transmits actual copy. Unlike computers, which transmit only files, fax transmits any graphic image including signatures, letterhead stationery and halftones. Unlike telex, which is slow and expensive, fax transmits vast amounts of information relatively quickly for about the price of a business phone call.

Fax works on a single international standard, which means that every machine can transmit to and receive from every other machine anywhere in the world. In fact, some countries rely more on fax than they do on telephones to send and receive information. For example, Japan currently has the largest installed base of fax users in the world. In Japan, to do business without a fax machine is like trying to play baseball without a mitt!

WHO SHOULD READ *THE BOOK OF FAX*

The Book of Fax will be of use to anyone interested in fax technology, particularly those who want to purchase a fax machine. *The Book of Fax* is written for nontechnical readers who need to learn the basics of fax equipment, and its use and management.

Because the authors and the publisher aren't affiliated with a manufacturer or other interest group, this book's impartial approach can help you objectively determine your needs—or whether you need fax technology at all.

For those who already own fax, *The Book of Fax* outlines ways to obtain the full benefit of fax technology in the most cost-effective way. Too often, when people buy new technical equipment, they use the easiest features and apply them only to existing ways of doing business. Because fax allows professionals to communicate in ways that weren't possible in pre-fax times, this book explores many new fax management techniques, including some important "dos and don'ts."

WHAT'S INSIDE

The Book of Fax is a complete guide to purchasing, using and managing fax. In it, you'll find

1. An introduction to facsimile technology—how it works and how it can help your business.

2. A step-by-step analysis to determine whether you need fax.

3. A comprehensive review of the major fax features—from the most common to the most advanced.

4. A systematic method to find the right fax to fit your communications needs and your budget.

5. A tested set of procedures for effectively and professionally managing fax, including some important "dos and don'ts."

6. A thorough overview of PC fax—and its pros and cons.

7. A resource guide to the fax community.

HOW TO USE *THE BOOK OF FAX*

If you don't yet own fax (or have just purchased one and aren't sure how it works), Chapters One through Three will be of particular interest. You'll learn about fax features and how to assess them to determine which are best for your business and budget.

> **Professionals can't afford to be fax-illiterate.**

If you already own fax, you may want to skim the first three chapters for updated information, then read Chapters Four through Eight for information on using and managing fax in your office or home. In each chapter, you'll find charts, worksheets or checklists that help you clarify needs and alternatives.

DON'T BE LEFT BEHIND

Fax is different from VCR's or telephone-answering machines. Professionals can't afford to be fax-illiterate. For many businesses, fax is as integral a part of communicating as the telephone.

As people discovered with telephone technology, it pays to stay informed. By being up-to-date you can make the best decisions and better communicate with other professionals, including consultants and sales personnel.

The Book of Fax is written to keep you at the forefront of this important new technology. After reading it, you'll be able to determine whether you need the strategic edge that fax can offer. You also will be able to better manage the communication channels you currently use. Let's get started.

Why Fax?

With all the talk about the paperless office of the future, why would anybody want to invest in a technology that generates more paper even faster? People are accustomed to picking up the telephone, using overnight delivery services, sending messengers and transmitting telexes. Until now those channels have sufficed, so why invest in a new technology like fax?

Like it or not, paper (or *hardcopy*, as the jargon goes) is here to stay, at least for the foreseeable future. In fact, the widespread use of personal computers has actually *increased* the amount of paper generated by most businesses. The only difference is that for many businesses to remain competitive, information now must be sent and received even faster.

> Fax provides a communications capability that no other alternative offers.

With its unique ability to transmit existing graphics as well as text instantaneously, fax provides a communications capability that no other alternative offers.

WHY NOW?

Twenty years ago, fax was an expensive, slow, smelly, noisy technology, used only by international stockbrokers who hid the machines in their supply rooms and used them only when absolutely necessary. However, recent technological breakthroughs have made fax faster, quieter and less expensive, allowing businesses to take advantage of its unique abilities.

> **More than four million fax machines will be purchased by 1991.**

New, improved fax was snapped up quickly by businesses that relied heavily on international communications, had branch offices or needed to transmit visual information over long distances. But so many new uses have emerged that fax is quickly becoming a mainstream technology.

According to researchers, more than four million fax machines will be purchased by 1991, and more than one million people will buy fax machines in the next twelve months. Yet many people still know very little about the pros and cons of fax.

MISCONCEPTIONS ABOUT FAX

Because fax makes duplicate copies of original documents, some people think of it as little more than a remote photocopier.

Because fax can receive incoming transmissions unattended, some people think of it as little more than a telephone-answering device.

Because most fax machines are easy to operate for their most basic procedures, some people think that fax is a simplified version of computer technology.

Finally, because fax provides instant transmission of documents, some people think of it as nothing more than a high-speed document delivery system.

Yes, fax makes duplicates of originals; yes, it can work unattended; yes, it transmits information electronically. But fax is more than a combination of existing technologies.

Fax is a strategic communications tool. Like all tools, it can be used and abused; it can work for or against you.

Tó get the most out of fax technology, you must understand both what it can and can't do. The challenge for fax users is not just to transmit information quickly, but to ensure that the right information gets to the right person at the right moment.

Whether you need fax equipment depends on how you use traditional methods of communication, the kind of information you deal with, its urgency and a host of other factors. By the end of this chapter, you should have a clear idea of your fax requirements, and how much time and money you might save by using fax. Chapters Two and Three describe basic and advanced fax features and help you match those features to your needs.

COMPARING COMMUNICATIONS OPTIONS

You'll better understand fax's advantages and disadvantages by comparing them with the traditional forms of communication you now use.

Most of the ways information is distributed are so routine that you seldom think about them. You intuitively know whether to write a letter or make a phone call. Yet in determining whether to add fax to your communications system, you should take time to review the advantages and disadvantages of existing channels of communication.

The following comparative analysis should clarify the kinds of information you now send through traditional channels that may be better suited for fax transmission. Although this may seem an obvious exercise, it will help you visualize the role of fax in your overall communications functions.

The Mail

ADVANTAGES:

Though much maligned, the good old US Postal Service offers several advantages.

1. Mail can transmit information in many different forms and

configurations, such as books, art and graphics, film and other media.

2. Mail provides complete coverage. It can be sent virtually anywhere in the world and requires no special equipment on the part of the recipient.

3. Mail is easy to use, and everybody understands how to send mail. Most companies have efficient routines for processing and routing mail.

4. Mail is extremely private. Both custom and law dictate that only the addressee and those authorized by the addressee can open a letter or package.

5. Mail is inexpensive.

DISADVANTAGES:
1. Mail is slower than most alternatives.
2. Delivery may be erratic and out of the sender's control.
3. If your recipient gets a lot of mail every day, your message must compete for attention.
4. It's difficult to verify whether a message has been received.
5. Mail is often time-consuming to produce and process.

The Telephone

In recent years, both technology and market competition have dramatically enhanced the usefulness of the telephone.

ADVANTAGES:
1. Telephone communication is instantaneous.
2. Telephone conversations are spontaneous and allow close interaction between participants.
3. Conversation is relatively private—both law and custom dictate that people not eavesdrop or make unauthorized recordings of telephone calls.

4. Like the mail, telephones provide very broad coverage—virtually everyone has a telephone.

> **You have no written verification of information communicated over the phone.**

5. Telephones are easy to use; little training is required.

DISADVANTAGES:
1. Telephone conversations require that both parties be available at the same time.
2. Except for short messages, information usually can't be transmitted through a third party for the recipient's later use.
3. You have no written verification of information communicated over the phone.
4. Because oral communication is a slow means of transmission, the telephone is often a poor medium for complex information (such as financial statements or engineering specifications).
5. Visual data, such as architectural plans, cannot be transmitted effectively over the telephone.
6. Employees can easily abuse the telephone by using it for personal use. Unmonitored telephones can generate substantial bills.

Overnight Delivery Service and Messengers

As the need for high-speed communication has proliferated, services have arisen to move data physically from one place to another—within a city, across the country or around the world.

ADVANTAGES:
1. Overnight delivery enjoys many of the same advantages as mail.
2. Overnight delivery is fast.
3. Unlike the telephone, senders can calculate their costs in advance.
4. Overnight express service is a more dramatic, urgent form

of communication than routine mail.

5. Local messenger services provide even better delivery schedules than do overnight services and convey an even greater sense of urgency.

DISADVANTAGES:

1. Overnight delivery (particularly for international delivery) and messenger services are expensive.
2. Although coverage is broad, overnight services don't reach all rural areas.
3. Messenger service usually is available only locally.
4. The sender is dependent upon pickup and delivery schedules controlled by a third party.

Telegram and Telex

Although domestic telegrams largely have been replaced by the telephone, telegrams still are used frequently for international business. The use of telex, on the other hand, has increased rapidly both domestically and internationally.

ADVANTAGES:

1. Both telegrams and telex are dramatic forms of communication that command immediate attention.
2. They can be sent at any time and the recipient need not be present.
3. Telex and telegrams have broad international coverage.

DISADVANTAGES:

1. Depending upon the volume of use, telex and telegrams can be very expensive.
2. Only small amounts of information can be sent in text form.
3. Telex and telegram messages are cumbersome to generate.
4. Telex and telegram messages are inconvenient to file.
5. Although you can know that the message arrived at its destination, you can't be sure that the person for whom it

was intended actually received it.

6. Telex isn't private. Anyone near the telex terminal can read transmissions.

Computer Networks

ADVANTAGES:

1. Transmission is immediate.
2. Information arrives in a form that can be revised easily.
3. Privacy and security systems can be built into the network.
4. The recipient need not be present.
5. The sender receives confirmation that the message has reached the intended destination.
6. Messages can be accessed quickly from remote locations.

DISADVANTAGES:

1. Coverage is limited to recipients who have appropriate equipment.
2. The technology is complex and costly.
3. Many computer networks cannot transmit graphics.
4. Standard communications protocols haven't been established, particularly for international communication.

> Fax technology combines some of the best and worst features of the traditional alternatives.

5. Considerable (and often expensive) training is necessary.

Fax

Fax technology combines some of the best and worst features of the alternatives described above.

ADVANTAGES:

1. Like the telephone, fax transmits instantaneously.
2. Like mail, fax can send both text and graphics of virtually any length and complexity.
3. Like telex and telegrams, fax transmissions are dramatic, conveying urgency and demanding attention.

4. The recipient need not be present to receive a fax message.
5. The sender receives written confirmation that the message has been received.
6. Costs are moderate, amounting to little more than the telephone bill.
7. Messages arrive in a form that can be filed easily.
8. Fax technology is relatively simple, requiring less training.

DISADVANTAGES:
1. Though fax use proliferates, many businesses still haven't adopted fax.
2. Though basic fax is easy to use, advanced features are sometimes complicated.
3. Fax is the only channel of communication for which the recipient and the sender incur supply and maintenance costs. Although most people buy fax to send information, they usually become recipients in short order.

> **Fax can complement traditional communications methods, saving you both time and money.**

4. Because sending costs are low, senders may transmit unsolicited fax messages demanding attention to a matter that the recipient might not wish to address.
5. Fax is so new that many organizations don't have routine procedures for handling incoming fax messages.
6. Fax messages aren't private; anyone at a fax terminal can read them.
7. The transmission quality of fax machines is uneven.

Some of the disadvantages (particularly the fact that some businesses don't have fax machines and the lack of internal procedures to handle fax traffic) will lessen as fax use grows. However, other disadvantages—such as costs to recipients and being at the mercy of unsolicited transmissions—will grow.

It's important to recognize that fax does not—and will not—replace the channels of communication you now use. However, for specific types of information, fax can complement traditional communications methods, saving you both time and money and giving you an important competitive edge.

UNDERSTANDING YOUR INFORMATION NEEDS

You can better evaluate the kinds of information with which you work by answering six important questions about your communication needs.

1. With what kind of information do you work? Is it primarily text (e.g., accounting reports, spreadsheets, raw data), or is it highly visual, containing many illustrations, photographs, charts or graphs?

2. How time-sensitive is the information you generate and receive? After you receive something, do you have three hours or 30 days to respond? Do you have difficulty generating information and receiving responses within the required deadlines?

3. What volume of information do you send or receive?

4. Who produces the information with which you work? In general, is your information generated by one person working with a computer or calculator (e.g., accounting reports, spreadsheets, correspondence), or do you work with groups of people to produce company annual reports, advertising and public relations, or strategic plans? In other words, is most of the information with which you work the product of an individual or the product of a group?

5. Where is most of your information generated? Do you work primarily with internally generated information (e.g., profit and loss statements, internal memos, minutes to meetings), or do you work with information generated outside your location (e.g., correspondence, bids, contracts, purchase orders, invoices)?

6. Who reviews and receives the information you generate? Is your information circulated primarily to other people at your workplace or to people outside your work location, such as remote offices and divisions, customers or suppliers? If your information generally is reviewed internally, where are the other people located who need to see what you produce?

HOW COST-EFFECTIVE IS FAX?

Your answers to the next series of questions can help you determine whether fax will be cost-effective for your business. Because you probably aren't familiar with all the fax features covered in Chapters Two and Three, your cost and savings estimate will be crude. However, by completing the section below, you should gain some sense of your fax needs and potential savings.

1. How much money do you spend per month, excluding equipment and administrative charges, on the following:
   a. Telephone usage $_____
   b. Postage $_____
   c. Telex and telegrams (include equipment and administrative charges in this figure) $_____
   d. Overnight delivery $_____
   e. Computer networks $_____

(Equipment and administrative costs are included in the cost for telex and telegrams because fax conceivably could be a cost-effective replacement for telex. However, it won't replace your telephone system or computer network. Indeed, it probably won't even reduce the number of stations you need.)

2. Give the percentage of information you send via each channel that could be better handled by fax transmission.
   a. Mail _____%
   b. Telephone _____%
   c. Telex and telegrams _____%
   d. Overnight delivery _____%
   e. Computer networks _____%

3. Now calculate the total savings, based on the percentages above.
   a. Telephone charges $_____

b. Equipment costs $_____
c. Administrative costs (outgoing) $__ __
d. Maintenance service and supplies $_____

4. Would there be additional administrative costs in handling incoming fax messages? In other words, what volume of fax traffic do you anticipate? How much time would that take? How much would that time cost?

By calculating the percentage you would save on your traditional communication costs and subtracting your anticipated administrative costs, you can determine your approximate budget for fax equipment right now.

> **FORMULA:**
> **Anticipated Savings - Anticipated Administrative Cost**
> **= Approximate Fax Equipment Budget**

Interpreting the Results

If your anticipated fax equipment budget is $500 or more, you might want to continue to investigate fax. But even if your potential savings are small, you can get an accurate picture of how fax might fit into your business by imagining what you could do with fax capability that you can't do now.

Fax is an evolutionary technology. Once fax is in place, people often discover new ways to put it to use. Consequently, in determining whether you need fax, consider not only what you can do with it today but what you might do with it in the future —a key consideration in determining which features you need.

Item: A small company sells most of its products to a large corporation. The large corporation often is slow to pay its bills. When the small company calls to collect, the accounts payable supervisor denies that they ever received the invoice. In response, the accounts receivable manager at the small company begins faxing a copy of the invoice immediately prior to calling. If the customer claims not to have a copy of invoice, the accounts receivable manager informs him that a copy of the invoice is in the customer's fax machine at that very moment, thus cutting down the time it takes to collect payment.

Item: A company finds that whenever its salespeople exhibit at trade shows, they're virtually incommunicado. Moreover, it's often inconvenient for the salespeople to send orders written on the show floor back to headquarters promptly. The company includes a fax machine in its booth display to send messages to its salespeople and to receive orders immediately.

Item: Before and during the process of writing proposals, a consultant likes to receive extensive input from the client. He often spends hours on the telephone thrashing out the details. He shortens the process by developing written material on his own and faxing that information to the client before each phone call. Because the client has time to consider each version beforehand, telephone calls are shorter, and fewer are needed to develop the final proposal.

MOVING ON

By now, you should have some idea of how you would benefit from fax. The next two chapters provide an overview of how fax works and its basic and advanced features.

2

A Look at Fax Features

FINDING FEATURES TO SUIT YOUR NEEDS

Understanding the key characteristics and features available in different machines is the next step in exploring the potential of fax technology.

Like other advanced technology, fax equipment is available in a wide price range. Stripped-down units can sell for less than $500. More sophisticated models cost more than $3,000. But price alone cannot be your major crite-rion. Low-end fax machines may not be suitable for the tasks you want to perform. On the other hand, the high-priced models may be needlessly complicated and packed with features you'll never use.

> Compatiblity is the most important fax feature.

Not all fax machines are equal, and not all are suitable for your business. The features of any fax may be roughly categorized as either basic or advanced.

Basic features are shared by all fax machines, from standard workhorse models to the most complex, state-of-the-art, multi-purpose equipment. In deciding which fax machine to buy or

> Like other advanced technology, fax equipment is available in a wide price range. Not all fax machines are equal, and not all are suitable for your business.

lease, you can only benefit from an understanding of the basic features.

Advanced features provide additional capabilities at additional cost. And though they sound wonderful in theory, many are still clumsy to use. Understanding the advanced features will help you determine the level of sophistication you need. Bear in mind that what might be a frivolous feature for one business can be vital for another.

BASIC FAX FEATURES

The first step in analyzing fax technology is to review the basic operational features that all fax machines share and that allow them to communicate with each other.

The most important feature is compatibility. Although fax technology enjoys internationally accepted communications standards, not all machines use the same standards.

Other key features to consider are the type of paper a machine uses, the size of documents it can send and receive, the quality or resolution of the image it can send and receive, its transmission speed, and the size and weight of the unit.

Compatibility

As with personal computers, compatibility (a machine's ability to send and receive material from another machine) is a crucial factor. Fortunately, international standards for fax technology have been established by the International Consultative Committee on Telegraphy and Telephony (CCITT), based in Geneva, Switzerland. As technology has improved, the CCITT has authorized improved communication standards as well.

Currently, four international standards exist for fax technology: Group 1, Group 2, Group 3 and Group 4 (which works with digital networks [ISDN]). The standards differ in the way information is scanned and coded and the speed at which it's transmitted and received. For most purposes, Group 3 machines are now the standard and are likely to serve your needs for many years.

Group 1 was the first standard approved by the CCITT and can be found on early fax machines. Group 1 transmission is extremely slow. Because the entire technology is outdated, don't even consider purchasing Group 1 machines.

Group 2 also employs an analog method but has faster transmission rates (about three minutes per page). Unless you have clients who use them, you should have no need for Group 2 machines.

In Group 3 technology, an image is scanned and the information is *digitally* encoded. In other words, each pixel, or dot, of the original is assigned a number. That number is then transmitted to the receiving machine, which prints the image corresponding to that number. The digital coding methods used in Group 3 fax technology are called *modified Huffman* and *read encoding*.

The most important breakthrough for fax technology has been the speed at which information is transmitted. Group 1 machines take six minutes to send or receive a one-page document that a Group 2 machine can handle in about three minutes. Group 3 machines take less than a minute to send the same document, and some take as little as 12 seconds.

Because fax standards for compatibility are agreed upon internationally, all machines within a group can communicate with each other. That means that all Group 3 machines can communicate with other Group 3 machines no matter where they may be.

However, not all machines can communicate between groups. For instance, a Group 1 fax cannot transmit to a Group 3 fax. And not all Group 3 fax machines can transmit to Group 1 and 2 models.

Most fax machines sold today use the Group 3 standard. But if you plan to send material to Group 2 machines as well, be sure that the machine you choose has Group 2 capability. Keep in mind that even if it takes your Group 3 fax only 20 seconds to send or receive a document, it still will take it six minutes to communicate with a Group 1 machine (provided your machine has Group 1 compatibility).

Furthermore, many advanced features found in Group 3 faxes cannot be used when communicating with machines of other groups. For example, Group 2 cannot receive documents with the same resolution and clarity as Group 3.

Despite the fact that all Group 3 machines can communicate with each other, they're not all the same. The CCITT established only minimum standards. Several manufacturers have added higher performance to their machines. For example, to be Group 3 compatible, a machine must be able to transmit information at 2,400 baud or 2,400 bits (a 1 or a 0) of information per second. Some Group 3 machines transmit at 9,600 baud or more. Moreover, although all Group 3 machines use modified Huffman or read to encode scanned information, many machines have optional alternative encoding systems as well.

Transmitting at higher baud rates and using alternative coding schemes increase transmission speed. Some of the features added to the minimum standards authorized by the CCITT are supported by all manufacturers; others are not. (Special features not included in the Group 3 standards are noted throughout this chapter.)

The Group 4 protocols are being developed to work on digital networks. However, ISDN networks won't be significant to most fax users for several years. Indeed, many telecommunications managers joke that ISDN stands for "I still don't need (it)." It's clear that Group 3 fax should serve your purposes for some time to come.

Resolution

As in photography, resolution refers to the detail with which a document or image can be reproduced by a fax machine. Basically, when it receives a document, a fax machine reconstructs the image of that document dot by dot by dot. The more dots per inch, the finer the resolution and the clearer the image.

| Normal transmission | Halftone transmission |

The Group 3 standard defines two levels of resolution—regular and fine. The type of resolution you use depends largely upon the amount of visual detail in your document. Regular resolution is best suited for text-based information and simple line drawings. Fine resolution is most appropriate for material with very small print and illustrations that contain fine detail.

> As laser printers and faxes become integrated, fax will gain higher resolution.

Regular resolution constructs images consisting of 203 horizontal lines per inch and 98 vertical lines per inch. That means each inch of the document is divided into 203 horizontal lines. Each inch of each horizontal line consists of 98 dots. In comparison, a modern computer laser printer has a resolution of 300 dots per inch. A good quality 12-inch-diagonal computer display has a resolution of about 100 dots.

This is the sentence in its original form.

This sentence was faxed in regular resolution.

This sentence was faxed in fine resolution.

In fine mode, the number of horizontal lines remains the same, but each inch contains 196 instead of 98 dots. Because each fine-resolution dot is half the size of a dot in regular resolution, an image consists of twice as much information and thus has significantly more detail.

Although it delivers more detail than most computer and television screens, fax still cannot produce the fine detail of a modern laser printer. As laser printers and faxes become integrated (and advanced Group 4 modes become available), fax will gain even higher resolution.

Very small documents that contain a great deal of detail may not be legible when transmitted in regular resolution. On the other hand, because it contains more information, fine resolution takes significantly longer to transmit.

Some manufacturers offer their own nonstandard resolutions to transmit documents that require an even greater degree of detail. However, these resolution schemes may not be compatible with other manufacturers' fax machines. In those cases, the machines can still communicate using the fine and regular resolutions.

Paper

Fax machines use two types of paper: thermal and regular. The thermal printing process requires thermal paper, which is coated with a colorless dye that responds to heat. Instead of striking the paper like a typewriter, the fax's print head heats the paper. As the paper is heated, the dye turns black, forming the image.

Most fax machines that use regular bond paper employ the same printing process as photocopy machines or a process called thermal transfer. A few use an ink jet process in which tiny drops of ink are sprayed on the paper. Still others use dot matrix printing, as do many computer printers. However, most fax machines that can handle bond paper use the xerographic process found in photocopy machines. Documents received on machines that use bond paper have the same look and feel as a photocopy.

Each type of paper has advantages and disadvantages. Thermal paper generally costs more. Because it's usually sold in rolls instead of sheets, the recipient must cut the message after receiving it or have a fax machine with an optional cutter feature. Thermal paper looks and feels less professional than normal paper. Finally, thermal paper tends to degrade under high heat. For example, a fax document left on the dashboard of a car on a hot summer day will turn completely black!

On the other hand, the thermal printing process requires no toner or drum (as does the photocopy process). Further, because a thermal printer has almost no moving parts, maintenance costs are significantly lower. Because the paper comes in rolls, thermal paper can record documents up to 39 inches long, a handy feature for faxing spreadsheets, large drawings, architectural plans or other nonstandard documents.

> **A fax document left on the dashboard of a car on a hot summer day will turn completely black!**

Bond paper offers two advantages. Documents on bond paper generally look better and resemble papers you work with every day. In addition, the paper costs less.

But bond paper has its limitations. The xerographic process requires that toner be replenished and the print drum be replaced periodically. Most bond paper comes in standard 8.5-inch by 11-inch sheets and cannot be used to receive documents longer than 11 inches in one piece. When long documents come in, single sheets must be taped together to record the complete image.

As fax technology is integrated into computer technology, fax documents eventually will be printed by the same printers your computers use. Chapter Six discusses this and other special features.

Document Size

When evaluating a fax machine, be sure that it can accommodate the page sizes you'll need. An engineer trying to transmit drawing plans has very different needs than an import/export service transmitting short memorandums.

Each fax machine has a minimum and maximum document size it can transmit. The most common minimum is 5 inches by

5.8 inches. Fax machines that have the same type of reader found on photocopy machines (in which you lift a cover and place the document flat on the glass) have no real size minimum. If you plan to send existing business forms via fax, the minimum document specification can be important.

Inequivalent machines designed mainly to send notes cannot accommodate full-sized letter paper, but most can transmit a document 8.5 inches wide, or the width of a standard sheet of paper. Some, however, cannot transmit information that runs from edge to edge. In other words, the scanner must have margins.

> **Because it affects your phone charges, transmission is a key feature.**

Unattended Operation

Virtually all machines can receive and print a fax automatically without the intervention of an operator.

However, if your machine shares a line with a telephone and doesn't have an auto voice/data switch, the fax machine won't be able to receive messages unattended during those periods of the day in which the telephone must be accessible.

Transmission Time

Because it affects your telephone charges for fax transmissions, transmission time (the length of time it takes a document to get from here to there) is a key feature and selling point. It also may affect your administrative costs associated with fax. Consequently, manufacturers often present impressive numbers to prospective buyers to convince them that their machine is the fastest. Unfortunately, the machine's performance in the workplace may not match the manufacturer's performance claim.

Your actual transmission time is influenced by several factors. Consider, for instance, the speed, or baud rate, of the modem (the device that actually sends the information over the telephone lines). As mentioned earlier, Group 3 fax machines send informa-

Transmitting and receiving at fine resolution require more time than transmitting and receiving at regular resolution.

tion at a minimum of 2,400 characters per second (or 2,400 baud).

Many machines have built-in modems that transmit at higher speeds—9,600 baud or more—but these high transmission rates can be used only for communicating with fax machines of equal speed. Although 9,600 baud seems to be emerging as the most common transmission rate, some low-cost personal fax machines transmit and receive at only 4,800 baud, or half speed. Other machines transmit at twice the speed of a 9,600-baud modem. In evaluating a fax machine, find out what baud rates are available for it.

Remember, fax machines must communicate at the same baud rate. In other words, a 9,600-baud machine communicating with a 4,800-baud machine will send at 4,800 baud. Some high-speed modems communicate only with fax machines made by the same manufacturer—and sometimes work only with the identical model from the same manufacturer. Unless you check these details before buying, you may be paying extra for features with limited practical application.

The second factor that influences your transmission time is the resolution of the image to be received. Transmitting and receiving at fine resolution require significantly more time (and telephone cost) than transmitting and receiving at regular resolution.

Several manufacturers include special techniques to speed up the rate of transmission. Some *compress* data, so that only meaningful information is transmitted. Data compression doesn't change what is printed, it merely changes how much information is transmitted over the phone line.

In general, however, compressed data must be *decompressed* at the other end, a feature that may not be available on the receiving machine. In addition, some machines provide for white *space skipping*; that is, areas of a document that have no image are scanned very quickly, speeding transmission time.

Bear in mind that unless both machines support the same

method of speeding transmission, the normal standard will be used. As fax becomes more popular and competitive, higher transmission

> The earliest fax machines were large, smelly devices, often hidden in the basement.

speeds and more efficient transmission features will become the norm. For now, it's important to remember that many of these features must be matched by the features of the receiving machine. Otherwise, a salesperson's promises about high-speed transmission may be technically accurate but of no value in the workplace.

Size and Weight

The earliest fax machines were large, smelly devices, often hidden in the basement. Today some desktop units weigh less than 10 pounds and take up little more desk space than a telephone-answering machine. Some full-featured models with large document feeders, however, can weigh as much as 40 pounds.

Because fax machines work with all telephone lines, some battery-operated models are designed for portability. Indeed, sending faxes from the back seat of a limousine is an emerging status symbol among the rich and frivolous. Auto faxes are becoming popular in the trucking industry, where signed bills of lading are a daily part of doing business. (In fact, many truck stops now offer fax service.)

ADVANCED FEATURES

Advanced features such as the following are available on machines that are more expensive than the basic models:

1. Transmission features accelerate the actual sending or receiving of a document.

2. Production features improve the quality of the document in hardcopy.

3. Management features contribute to managing fax technology more efficiently.

ADVANCED TRANSMISSION FEATURES

Not surprisingly, because fax is a communications tool—not a production tool—most advanced features are designed to improve and streamline transmission.

Built-in Dial

Although fax machines use regular telephone lines to transmit and receive documents, not all fax machines have telephone-calling capability. With the simplest fax machines, a separate telephone handset must be plugged into the same phone jack as the fax. To send a fax, the user must dial a number by using the telephone and then direct the fax machine to send the document.

Built-in dial capability is perhaps the most commonly available advanced feature. With a built-in dial, a user can dial a number directly from the fax machine, eliminating the use of a separate telephone.

Although convenient, a built-in dial cannot be used if a person, not a fax machine, is at the receiving end. Organizations often don't reserve a telephone line specifically for fax usage, because incoming fax traffic may not be heavy enough to justify a dedicated line. Moreover, at some companies all incoming calls may be routed through a switchboard or PBX.

In these situations, called *manual receive*, a person answers the call and then turns on the fax machine. If you frequently send fax messages to companies using manual receive, you must be sure that your machine includes not just a built-in dial but a complete telephone handset integrated into the unit so that you can talk to the person at the other end and ask them to activate their fax machine.

As an economical alternative, you can add this feature to any fax machine you buy simply by plugging an inexpensive telephone into the same telephone jack.

One-touch Dialing

Fax machines with built-in dialers often offer alternatives to automate and accelerate the speed at which another fax machine can be dialed: *one-touch dialing* and *speed dialing*. As with memory telephones, frequently used telephone numbers can be programmed and stored, then dialed by pushing the button assigned to each. One-touch dialing is a particularly useful feature if you plan to send faxes regularly to a specified group of recipients, such as a small network of field offices or a few major customers.

If your telephone system requires that a number be dialed to access an outside line, the fax machine's one-touch dialing must have the ability to *pause* between digits. Many machines include a pause key or a special character that produces a two-second delay. Consequently, to program the fax machine to dial the number 555-1212 in a telephone network that requires the number 9 to be dialed to access an outside line, the user would enter 9-pause-5-5-5-1-2-1-2.

In analyzing a fax machine's one-touch capability, be sure that each memory can hold a sufficient number of digits to meet your requirements. Dialing a long distance number within the US requires at least 11 digits, perhaps more, depending upon the requirements of your telephone system. Dialing an international number can require 14 or more digits.

Speed Dialing

Speed dialing isn't quite as fast as one-touch dialing but is much faster than dialing the complete fax number each time. In essence, speed dialing allows telephone numbers to be stored and accessed using a two-digit code. As a result, you can place a call with four keystrokes: one stroke to access the speed-dialing feature, two strokes to access the number, one stroke to initiate the dialing process.

> **Speed dialing helps to avoid misdialing.**

Many telephone users have found speed dialing a frivolous feature. The time you save by reducing the number of keystrokes is more than offset by the time consumed programming speed-dialing numbers and then looking them up. However, with fax, speed dialing can be a useful technology.

First, most companies these days have separate fax numbers and few people bother to memorize another company's fax number. Nor do they dial the number so frequently that they learn it by rote. (Indeed, many executives don't even know their own fax number; when asked, they have to check their business cards!) As a result, most fax operators refer to a directory for each number they must dial. By keeping a directory of your 30, 50 or 100 most commonly used numbers near the fax machine, the speed-dialing feature can be utilized efficiently.

Second, speed dialing helps to avoid misdialing. It's extremely annoying for a person to answer the telephone only to hear the squeal of a fax machine on the other end; and when dialing internationally, incorrectly dialed numbers are costly.

Most fax machines with speed-dial capability allow a name field to be associated with each stored fax number. For example, you can program the word *lawyer* to represent your attorney's fax number. To dial that number, you would look up the word "lawyer" in your directory and enter the speed-dial number associated with it (e.g., 02).

> Sending sensitive information to the wrong fax machine can be embarrassing, if not dangerous.

Some machines display the number dialed and/or the word "lawyer" on an LCD screen. A stored number directory that includes character names can be useful, because it eliminates the possibility of accessing the wrong fax machine—by dialing 02, for example, instead of 03.

Sending sensitive information to the wrong fax machine can be embarrassing, if not dangerous. But entering the names into a stored number directory has several drawbacks. Most fax machines don't have typewriter-like keyboards. Instead, they have numbers that can be used to represent letters. For example, registering the name "First City Bank" would require the following sequence.

```
16-19-29-31-32-10-13-19-32-37-10-12-11-25-22
 F  I  R  S  T     C  I  T  Y     B  A  N  K
```

Although not difficult, it is laborious and time-consuming. Alternative methods devised by some fax machine manufacturers for programming names are just as cumbersome.

Moreover, in some fax models, not enough letters can be stored in the directory to spell out the full name of the party to be dialed. For example, First City Bank requires 15 characters of storage (including the spaces), too many for some machines.

Automatic Redialing

Because fax machines use standard telephone transmission lines, your machine will get a busy signal if the receiving fax machine is already sending or receiving information.

Automatic redialing instructs your machine to continue to try to reach the same fax number at periodic intervals. With some models, the machine can be programmed to keep trying until it succeeds. Others will stop after redialing two or three times.

The usefulness of this feature depends upon how often you anticipate getting busy signals when accessing other fax machines. Automatic redial allows you to leave your machine unmonitored, even if the receiving machine is very busy.

Alternate-number Dialing

As fax technology becomes more common, busy signals will increase. Many companies will have several fax machines, each plugged into its own dedicated line. A fax machine with *alternate-number dialing* can be programmed to dial another telephone number automatically if the original number dialed is engaged.

Most machines with alternate-number dialing also have automatic-number redialing. Thus, if both numbers for the fax machine dials are engaged, it can be programmed to keep dialing—alternating between the two numbers—until they connect.

> **Automatic redial lets you leave your machine unmonitored.**

Of course, spending several hundred dollars for automatic redial and alternate-number dialing may not make sense if an operator is monitoring the machine. In that situation, the operator will know when a receiving machine is busy and can redial later. But if you wish to leave the machine unattended, both automatic redial and alternate-number dialing are important features.

Delayed Transmission

Both automatic redial and alternate-number dialing are useful for delayed transmission, a feature that programs the fax machine to send a document at a specific time—perhaps when it's unattended by an operator. (See Chapter Three for a detailed discussion.) In those situations, if the sending machine receives a busy signal, it must be able to continue dialing the same number or switch to alternate numbers until the document is successfully sent.

Automatic Voice/Data Switch

In many cases, a fax machine is plugged into the telephone network at a station where a telephone also is used. Indeed, with compact models, complete with integrated telephone handsets and answering machines, some fax machines are positioned to serve as a multifunction communications center.

An *automatic voice/data switch* allows the fax to determine whether the caller at the other end is a person or another fax machine. If the caller is a person, the telephone in the receiving unit rings. If the caller is a fax machine, the unit accepts the fax transmission automatically, without ringing the telephone. Many small businesses don't want to dedicate a line exclusively to fax use. If the fax machine must share a line with a telephone, automatic voice/data switching is essential.

> **If a voice/data switch is important to you, be sure to look beyond the claims.**

Automatic voice/data switching allows you to direct a fax transmission to a specific telephone number. For example, if a company has three incoming telephone lines, a fax with a voice/data switch can be plugged into the third line and that number can be publicized as the company's fax number. If a call comes in on that line, a machine with automatic voice/data switching senses whether it's a fax or a telephone call and responds accordingly.

Many voice/data switches differ from manufacturer to manufacturer. If this feature is important to you, be sure to look beyond the claims and determine exactly how they work. For example, one switch greets each caller with a screeching tone. After several seconds, it determines that the caller is, in fact, human, then rings the telephone. However, by that time, any sane person will have hung up.

> **Although low-tech, the document feeder is one of a fax machine's most useful features.**

In assessing this feature, it's extremely important to check with your telephone equipment supplier to make sure the fax machine is compatible with your telephone system. (For more information see Chapter Four.)

This feature is also available as a separate device that connects both your telephone and fax into the telephone line. Available for less than $100, this add-on device may cost you less than buying a fax that includes the feature.

Document Feeder

Although rather low-tech, the *document feeder* is one of a fax machine's most useful features. The document feeder inserts the original document into the fax machine one page at a time and eliminates the need for manual insertion of each sheet. Although having a document feeder doesn't increase the number of pages that can be sent, the disadvantages of manual feeding long documents are obvious.

Document feeders are essential for delayed transmission (see Chapter Three) so that the document to be sent can be placed in the document feeder to await the specified transmission time. In this situation (as in the case of unattended transmission), the capacity of the document feeder will determine the utility of the fax machine. Most commonly, document feeders can hold 5, 10 or 30 sheets of paper. A five-page document feeder restricts you to sending a maximum of two faxes at a time, because even a one-page fax takes two pages (a transmittal sheet and the fax itself).

Instead of using a document feeder, some fax models work more like copiers. The original is placed on a flat piece of glass and scanned. While this method is as inconvenient for sending long documents as it is for photocopying them, it's the only way to fax documents that are too cumbersome for a document feeder, such as pages from a book. A library, for example, would be well advised to consider such a model.

Automatic Paper-cutting

Many fax machines use paper that comes in rolls rather than sheets. This allows the machine to receive documents of a specific width but of variable length. However, when a multipage document is received, pages are printed one after the other on a continuous length of paper.

With *automatic paper-cutting*, each page is automatically cut from the roll after it's received. Pages of variable length can still be received, but each page is physically separated from the next. This feature eliminates the annoying necessity of cutting paper manually.

Copier Option

Because fax machines reproduce original documents, many manufacturers have begun to include a *copier function* on their fax machines. In copier mode, the fax machine operates like a simple office copier, making photocopies in all the resolutions the machine can support.

> The copier function offers you a rough preview of how a document will look when received.

For small businesses and individuals who make only a few copies, having a copier mode on the fax machine can be an attractive alternative to investing in a separate copier. However, there are limitations. Unlike copiers, fax machines only can make one copy at a time. Consequently, to make five copies of a document, the original must be fed through the machine five times. (The exception is those machines with internal memory.)

Moreover, it takes a fax machine about 30 seconds to reproduce an original, which is extremely slow. And the paper needed for fax copies costs substantially more than regular bond copy paper.

Nevertheless, the copier function can be an attractive feature, even if you don't need office copies. The copier function offers you a rough preview of how a document will look when received.

> **Fax technology quickly is becoming integrated with copier technology.**

This is particularly useful if you send complex documents with many details. Also, the copier function lets you experiment with fine mode, regular mode and halftone and allows you to decide which is the most appropriate prior to actual transmission. Because this reduces the number of times that documents must be retransmitted, it cuts costs.

Fax technology quickly is becoming integrated with copier technology in larger multipurpose machines designed to serve as the primary copiers in the office. With integrated fax/copiers, your copying needs must be assessed independently of your fax needs, and the machine must be able to meet both requirements.

Automatic Fallback

As mentioned earlier, fax machines send information over the telephones via modems, which send information at different rates. Most fax machines send information at high speeds (baud rates). Unfortunately, the faster the information is sent, the more likely that errors in transmission will occur and that some information will get lost or garbled due to "noise" on the telephone lines. As a result, portions of the document may be unreadable when received.

Fax machines with *automatic fallback* can detect the occurrence of errors and will begin to transmit at a slower rate. For example, many Group 3 fax machines transmit at 9,600 baud. If a machine with automatic fallback detects errors, the machine shifts to 4,800 baud or even 2,400 baud. Though the transmission time is longer, the document received will be readable.

Automatic fallback is particularly important for transmission overseas and to remote places. If you cannot hear somebody on the other end of a telephone line because of "dirty" transmission, chances are good that the fax machine cannot "hear" either, and automatic fallback may be a valuable feature.

Transmit-terminal Identification

In essence, *transmit-terminal identification* stamps each page of a fax document as it's sent with your fax telephone number, including the date and time of transmission and the page number. Some fax machines also can insert phrases—such as the name of your company—in the heading. This

> **Page identification helps the receiver collate the incoming document.**

identification is useful for fax transmissions that are several pages long. However, all fax transmissions should be accompanied by a cover sheet that includes the information listed above and the number of pages in the document.

MODE	CONNECTION TEL	CONNECTION ID	START TIME	USAGE T.	PAGES
TX	16883909	CH PRINTING	08/30 12:37	00'49	01(00)

```
                    RECEIVE  REPORT
                         JUN-26-88 SUN   8:44
TERMINAL  #
```

#	DATE	S. T.	NAME	TIME	PGS	NOTE	DP
01	JUN-26	8:37	0	2' 6"	1	OK	
TOTAL				2M 6S	1		
GRAND TOTAL				2M 6S	1		

Many fax machines stack incoming messages in order, face up. Therefore, the last page transmitted is on the top of the pile. Individual page identification helps the receiver collate the incoming document and recollate it should there be an accident. Moreover, having page numbers that correspond to the fax document, and not necessarily the original, can be extremely helpful, particularly if you're sending parts of an original that aren't numbered consecutively.

Transmit-terminal identification should be viewed as a convenience for the receiving party and insurance that your document will arrive intact at its destination, particularly if it must pass through several hands at the receiving end before it reaches the party for whom it's intended.

Broadcast Transmission

Broadcast transmission uses a combination of internal memory and stored fax numbers to "broadcast," or transmit, a document stored in memory to those fax numbers contained on the broadcast list. For example, price updates can be broadcast to hundreds of sales offices or customers every weekend after a single copy of the updates is scanned into the central machine. Then, each number on the broadcast list is called and sent the information.

Without this feature, a separate copy of the document would have to be loaded into the document feeder for each location, an approach that's affected by document-feeder limitations. If your machine serves as a central hub for information, this is a useful feature.

Many manufacturers claim broadcast capability even on their low-end machines. Reading the fine print, however, reveals that they do have that capability only when used in conjunction with an expensive machine made by the same manufacturer. Using that method, a document is scanned into the inexpensive machine, then is sent to the more expensive "hub" machine for true rebroadcast.

> Polling allows a fax machine to call another to request a document rather than send one.

If you need broadcast capability, be sure the machine you buy is able to broadcast directly. If the machine doesn't contain its own internal memory, it can't perform the broadcast function.

Polling

One of the most misunderstood fax features, *polling* allows a fax machine to call another to request a document rather than send one. Polling also lets you place documents in the fax machine that will be sent only at the request of a calling machine.

For example, a supplier of fine china has 100 independent salespeople selling its wares door to door. Each salesperson has a fax machine and at the end of each workday, the manufacturer receives a sales report.

Without polling, each of the 100 salespeople would have to call the supplier's fax machine. With 100 incoming calls beginning after the workday ends, the entire activity would have to be well-coordinated or it would take several frustrating hours to complete. With polling, the salespeople place their sales reports in their fax machines but don't send them. Instead, the manufacturer's fax machine automatically calls each salesperson's machine in order and requests the sales report. With automatic speed dialing, the procedure can be quick and efficient and requires no operator intervention.

Both the transmitting and receiving machines must be prepared in advance to poll (request documents from others) or to be polled (send documents on demand). To poll, a person simply loads a document into the fax machine and presses the poll function. In its basic form, any machine can call the machine that has been set up to poll and request the document. This is called *free polling*.

> **With free polling, the party to be polled cannot control who receives the information.**

In some situations, however, free polling can be dangerous. Suppose, for example, that a builder wants to poll several potential vendors for quotes on building materials. With free polling, an unethical competitor could simply request a competitor's quote. Indeed, with free polling, the party to be polled cannot control who receives the information, which brings up the subject of security.

Security

With the above situation in mind, several manufacturers include various levels of *security* with their polling function. The most rudimentary secure polling procedure requires that you enter a list of the telephone numbers for all fax machines that have polling privileges. Although this system limits the number of outsiders who can poll the machine, it doesn't ensure that information left in the fax to be polled won't be picked up inadvertently by an inappropriate recipient with polling privilege.

A second level of security requires not only that the telephone number of the machine requesting the document be listed, but that the terminal identification number also match before the document is sent. This ensures that only the authorized machine can poll your machine for a specific document.

For the most advanced level of security, not only must the telephone number and the terminal identification number be known, but the polling machine also must transmit a special security code. Consequently, the number of people who can poll your machine is limited. Unfortunately, this level of security generally works only with machines built by the same manufacturer. Consequently, if you plan to send information that needs sophisticated security, you may be restricted to one manufacturer as your requirement for fax equipment grows.

If you plan to gather information routinely from a large number of sources, polling is an essential feature. In such situations, polling manages the flow of fax traffic. Because the machine has the responsibility to contact each source of information, the need for human intervention is virtually eliminated.

Internal Storage Memory

The new generation of fax technology has increasingly come to resemble computer technology. Like computers, the most up-to-date fax machines can store documents electronically in their memory. *Internal memory* allows the fax machine to scan, document | **Store and forward can help reduce telephone charges.** |

and store information that can be "broadcast" to several other fax machines either immediately or at a later time.

Store and Forward

Store and forward combines memory capability with broadcasting. A machine can be programmed to receive a document, store it in memory and then retransmit it to other machines. Store and forward can help reduce telephone charges, particularly for companies with large fax networks spread across the country or the globe. For example, the headquarters of a large corporation

> **To take advantage of lower telephone rates, it's often more convenient to send a fax transmission at night.**

can broadcast the details of a personnel change to several regional headquarters, which in turn rebroadcast the same information to the local branches. Or a factory located in Asia can send its production figures for the day to a single US location, which rebroadcasts them to regional centers.

Delayed Transmission and Timer Transmission

To take advantage of lower telephone rates or big time-zone spreads, it's often more convenient to send a fax transmission at night. *Delayed transmission* and *timer transmission* allow specified documents to be sent to specific fax machines at specific times. Delayed transmission is usually limited to machines with either document feeders or internal storage memory. Delayed transmission offers perhaps the greatest potential for cutting fax costs. In practice, however, it's often ignored because it involves mastering a complex procedure.

Despite the learning curve involved in effectively using delayed and timer transmission, it may be worth the investment. For example, AT&T rates for transmitting from Los Angeles to New York drop by about 50 percent after 11 pm. If a five-page report is to be sent from Los Angeles to New York every day, delayed transmission can save you hundreds of dollars a year.

However, the usefulness of the delayed transmission feature depends on several factors. First, only documents that can be stored in the sheet feeder or in the internal memory can be transmitted on a delayed basis. If a lot of material must be transmitted every night, a machine with an internal memory is probably your best option.

Second, most machines can be programmed to send only one delayed transmission to one location at a time. If you plan to send material to several locations, you must be certain that the machine has that capability.

Programming for delayed transmission to different locations can be complex. Either you must enter the exact number of pages of each document to be sent to each location, or you must prepare specially coded sheets, called optical mark reader sheets (see below), to signal the machine when one document has ended and another is beginning. If your fax traffic is high, delayed transmission can keep your fax working 24 hours a day and is worth investigating.

Optical Mark Reader Sheets

Optical mark reader (OMR) sheets are most familiar in the form of the answer sheets for standardized tests—in which you fill in the appropriate circle with a No. 2 pencil. Some fax machines can read OMR sheets that contain information directing them to send a specific document to a specific location at a specific time.

In general, OMR sheets are used for unattended broadcast transmissions and delayed broadcasts. Each document to be sent is loaded into the document feeder with an OMR sheet detailing the instructions for the fax machine. Most machines require that the destination number already be stored in memory. The machine waits until the time specified on the OMR sheet, then sends the document.

OMR sheets aren't the only way to program unattended transmissions of multiple documents. Some machines can be programmed through their control panel in much the same way that a VCR can be programmed to record television programs at different times.

ADVANCED PRODUCTION FEATURES

Advanced production features are designed to enhance the appearance of the document so that the recipient's copy is as close to the original as possible. As with photocopy technology several years ago, the quality of the copy received is one of the most widely discussed issues in fax technology. Not surprisingly, production quality has become a key selling point.

Contrast Control

In many of its operations, a fax machine is like a copier machine; after all, it makes copies of an original. *Contrast control* helps the machine to improve the quality of the copy when the original document is either too light or too dark.

Fax machines are available with either manual or automatic contrast controls. With manual contrast control, if the original document is too light or dark or has too little contrast, you simply adjust the contrast control accordingly. Most models carry only three contrast settings: light, dark and regular.

Machines with automatic contrast control sense the quality of the original and make the necessary adjustments. If you frequently fax complicated, multicolored documents, this feature can be particularly useful.

> **Advanced production features enhance the appearance of the document so that the recipient's copy is as close to the original as possible.**

You can check the quality of the contrast control by operating the fax in copier mode. Experiment with high, low and regular contrast. Keep in mind, however, that telephone line noise often erodes legibility, so the results from the copier will be a "best-case" preview of how the fax will appear on the receiver's end.

Oversized Document Reduction

Those who plan to send spreadsheets printed on 11 by 14 7/8 inch computer paper or an engineering drawing on 11 by 17 inch paper will be particularly interested in reducing such oversized documents. The scanner size determines the largest document that can be reduced, usually 11 inches wide. The *oversized docu-*

ment reduction feature allows large documents to be received on standard-size paper with all the information intact, albeit smaller.

Fax machines won't print out large documents; they print them in

Those who plan to send spreadsheets or an engineering drawing will be particularly interested in reducing such oversized documents.

the reduced size. So, if you own a reduction copier and don't need to send a large volume of oversized documents, this feature isn't necessary.

On the other hand, if you work regularly with oversized documents, this feature may be more convenient than the extra reduction step.

Halftone Transmission

Most black and white photographs are made up of various shades of gray, or halftones. *Halftone transmission* allows a fax machine to record and reproduce subtly shaded photographs and graphics more accurately.

Halftone transmission is a feature determined almost exclusively by the transmitting machine. That is, if your machine has halftone capability, it can send halftone images to any Group 3 machine. The only exceptions are receiving machines with paper too insensitive to record the more subtle signals.

Halftone transmission generally is rated according to the number of different shades of gray it can transmit. Fax machines are available with 8, 16 and even 64 shades of gray.

If you intend to transmit a high volume of photographs and graphic material, test the halftone transmission capabilities *over real telephone lines*. The copier function can't provide a realistic test of the machine's capability. Because halftone transmission entails sending more complex information, the transmission is more susceptible to telephone line noise and interference.

RS 232 Port or Interface

Familiar to computer users, the *RS 232 port* is essentially a plug that allows two machines to be joined by a cable and pass information back and forth.

An RS 232 port on a fax machine opens up several options. First, it allows the fax machine to serve as a printer for a computer. Unfortunately, because most fax machines use thermal paper, they're much more expensive to operate than a standard computer printer—and the quality of the work is lower. Furthermore, most fax machines won't print graphic images sent by a computer. Finally, an RS 232 port probably will add more to the cost of the fax machine than the price of a good dot matrix printer.

> **The communications journal provides an audit trail of all fax activity.**

On the other hand, an RS 232 port may be a sound investment for both the present and the future. Some fax machines with an RS 232 port can be used to replace telex terminals. Telexes sent through personal computers can be received on the fax machine. This combination of technology eases the transition from telex use to fax use.

ADVANCED MANAGEMENT FEATURES

The final step in evaluating a fax machine—one that can prove costly in the long run if ignored—is the ease with which its use can be monitored and controlled. Reporting features are important to ensure efficient fax management.

Transmission Report

After a fax is sent, a *transmission report* records the telephone number of the receiving machine (and sometimes its terminal identification number, too), the number of pages sent, the transmission time and an indication of any errors in transmission. Attaching a transmission report to the original document is an efficient way of recording who has received a document and when.

Some expensive fax machines stamp the transmission report on the original, but this can be confusing if the document is to be retransmitted elsewhere at a later date.

Communications Journal

The *communications journal* provides an audit trail of all fax activity. It lists the date, time, duration, length and destination or source of all fax transmissions and receipts. The report can be printed on demand or after a full page of information has been accumulated.

The communications journal is a key management tool, indicating whether the fax machine is being used according to established company guidelines and whether faxes are being sent at the most economical times and to the right sources. It offers a cross-check for any fax logs you might maintain and a safeguard against unauthorized use.

Finally, for companies that bill clients for the faxes they send, the communications journal is a useful aid in preparing invoices.

```
*****************************************************************
*                                                       P.01    *
*                       SEND  REPORT                            *
*  TERMINAL #   001                      JUN-25-88 SAT 16:49    *
*                                                               *
*  NO. DATE   START      RECEIVER      TX TIME  PAGES  NOTE     *
*                                                               *
*   1 MAY-27   6:47 3518256               48"     1     OK      *
*   2          7:12 12134704790         4'35"     2     OK      *
*   3 MAY-31  17:41 3518256               58"     1     OK      *
*   4 JUN-16  16:19 3518256             2'50"     1     OK      *
*   5 JUN-18  11:17 3518256               52"     1     OK      *
*   6         11:22 3518256               47"     1     OK      *
*   7 JUN-21  16:00 3518256               55"     1     OK      *
*   8 JUN-25  16:35 3519698             1'08"     1     OK      *
*   9         16:36 3519698             2'08"     1     OK      *
*  10         16:40 3519698             2'08"     1     OK      *
*                                                               *
*                       TOTAL           17'09"   11            *
*                                                               *
*            GRAND TOTAL    TIME:     36M 48S   PAGES:    22    *
*                                                               *
*****************************************************************
.................................................................
*****************************************************************
*                                                       P.02    *
*                      RECEIVE REPORT                           *
*  TERMINAL #   001                      JUN-25-88 SAT 16:49    *
*                                                               *
*  NO. DATE   START      SENDER        RX TIME  PAGES  NOTE     *
*                                                               *
*   1 MAY-27   6:49 3518256             1'21"     1     OK      *
*   2          9:30 3518256               52"     1     OK      *
*   3 MAY-28   6:29 3518256               52"     1     OK      *
*   4 MAY-29   3:49                     **'**"     0    CANCEL   *
*   5 JUN- 6  11:46                     **'**"     0    CANCEL   *
*   6 JUN-16  15:18                        19"     0    CANCEL   *
*                                                               *
*                       TOTAL           3'24"     3            *
*                                                               *
*            GRAND TOTAL    TIME:     11M 27S   PAGES:    10    *
*                                                               *
*****************************************************************
```

MOVING ON

Although advanced features are packaged in various combinations, they can be effectively analyzed in three areas: transmission features, production features and management features. The advantages of each group depend entirely on how you plan to use fax. In some situations, the halftone transmission feature is irrelevant; in others, it's essential. Your goal, and the goal of the next chapter, is to match features to your needs and budget.

3

Matching Fax to Your Needs

CHOOSING THE RIGHT FAX

When selecting a fax machine be ever-mindful of the age-old axiom, get the right tool for the right job. Chapter One clarified the ways in which fax communication can be used. Chapter Two introduced you to the major features available today. This chapter helps you select a model that not only fits your budget but is equipped with the features that will meet your needs.

Determining which fax is best for you involves three steps:

1. Review the features you need.

2. Determine which combination of features (and prices) are available. For example, if you wanted an automobile with a leather interior, you might have to buy the mag wheels as well. So, too, in fax technology, features come in clusters, and those clusters command different price levels.

3. Choose an appropriate vendor from whom to buy the machine.

PRICE VS. FUNCTION

In the long run, the most important consideration in choosing a fax machine is function—how will the machine you buy enhance your business communications? But prices vary widely, so cost considerations can't be overlooked. In fact, price is perhaps more important in choosing fax equipment than in choosing other kinds of office machines. If you select a fax that pays for itself quickly, you haven't made a mistake. But you may save more money by buying a machine with additional features that further enhance your communications.

> As with most new technology, fax machine prices are plummeting.

However, cost isn't the only criterion for evaluating a fax machine. Because Group 3 compatibility promises to be the most important standard for several years, the fax machine you buy today will remain useful for years to come.

This chapter follows the three-step selection process outlined above. It begins with a review of the major fax features. A checklist is provided so that you can more easily determine which features best suit your needs.

Following the review is a general analysis of what features are available at various price points. As with most new technology, fax machine prices are plummeting, and new models with new combinations of features are being introduced. It's important to check the market for the latest advances.

Finally, when you know which features you need and what price you want to pay, you must decide from whom to buy. Fax machines are sold through various distribution channels, from retail stores to mail order companies. The benefits and drawbacks of each distribution channel are addressed in the final section of this chapter.

ASSESSING THE FEATURES
Compatibility

Compatibility is perhaps the easiest feature to assess. Don't settle for anything less than Group 3 compatibility. You may see small advertisements in your daily newspaper touting faxes "new out of the box" for less than $500. Be careful. Those machines often are only Group 2 compatible. And although many Group 3 machines can communicate with them, Group 2 machines are awkward and costly to use.

Bear in mind that Group 2 machines require three minutes to send or receive *a single page*. Hence, they tie up the telephone for long, costly periods of time.

Group 3 compatibility is essential. However, you also may need Group 2 capability if the people with whom you hope to communicate have Group 2 machines. If you expect to be a heavy fax user, survey the people with whom you plan to communicate. If all have Group 3 machines, Group 2 compatibility is superfluous. If some have only Group 2 compatibility, then your machine also should include Group 2 compatibility.

Group 1 machines are by and large obsolete, and you shouldn't consider paying a premium to add Group 1 capability

> Often overlooked, a fax machine's paper capacity is one of the most important keys to its utility.

to your unit. Group 1 requires six minutes to send or receive a page. So, if you have an important client who has only a Group 1 machine, it might pay to buy him or her an inexpensive Group 3 machine. You probably will recoup your costs in lower telephone charges and administrative overhead, and your customer is sure to be grateful.

Notes:
Number of Group 2 Receivers (if any) _____
Number of Group 3 Receivers _____

Receiving Paper Capacity

Often overlooked, a fax machine's paper capacity is one of the most important keys to its utility.

If you generate a lot of fax traffic, paper capacity determines how long the machine can receive messages without operator intervention.

For example, many law offices have begun to fax all but their most sensitive documents. A typical legal brief easily can be 50 or more legal-size pages. If the receiving fax machine has a 30-meter (98-foot) paper roll, it can receive only two such documents unattended.

There are three standard sizes for paper capacity.

Roll Size		Approximate Number of Standard Sheets
30 meters	(98 feet)	90
50 meters	(162 feet)	150
100 meters	(393 feet)	300

If you expect to receive a number of faxes or several large documents at a time, a large paper capacity is obviously essential.

Several factors tend to reduce the amount of paper available for actual fax messages. Fax protocol dictates that every fax message be sent with a separate transmittal sheet. Most machines send one or more lines of terminal identification information per page, adding slightly to the amount of paper needed to print a page. Many machines sending documents also transmit a reception report, printed out after each fax is received. This adds a few more inches of paper to each fax received. Receiving machines also print a separate reception report after a fax message is completed.

Many manufacturers signal that the paper roll is nearing the end by placing a red stripe down one side of the paper. The unsightly stripe can run for several feet. If you opt to change rolls as soon as the stripe appears, you further reduce the usable length of the paper.

Clearly, attention to paper capacity is critical. Do you need large paper rolls or paper bin capacity?

> A document feeder is essential if you plan to send multiple-page documents.

You may not. If your machine will be attended most of the time, or you plan to send faxes rather than receive them, paper capacity may not have much bearing on your choice of features.

Document Feeder Size

As with computer printers, document feeders can add significantly to the cost of the machine. Consequently, you should first decide whether you need a document feeder at all.

Infrequent users may be more willing to feed each sheet into the machine manually. Moreover, if the fax machine will be used primarily as a messenger service, shuttling short notes back and forth at a moment's notice, a document feeder may be superfluous.

A document feeder is essential if you plan to send multiple-page documents or send documents unattended. In either case, consider the number of documents you plan to send.

A 10- or 12-page report requires several minutes to fax. And time never seems to pass more slowly than when you're standing around watching a fax machine work. As with photocopying, if you find yourself daydreaming and don't feed the documents smoothly, you can make costly mistakes.

> **If you have access to a reduction copier, you may not need scanner capability.**

Sheet feeders and internal memory are required to utilize delayed transmission and polling. Both features, in the long run, can generate significant savings in telephone costs.

Sheet feeders commonly come with a capacity for 5, 10 or 30 pages. Be sure the feeder can handle your normal daily volume. If you plan to send only three faxes of five pages using delayed transmission, your sheet feeder must have a minimum capacity for 15 pages. And it's wise to plan for growth as well.

Notes:
Plan to use delayed transmission or polling? Yes____ No____
Number of pages of average fax message _____
Document feeder capacity needed _____

Telephone Handset

You'll probably want a fax machine with a telephone handset. However, because the handset either can be integrated in the fax unit itself or simply plugged into the same telephone jack in the wall, you can view the handset as an optional feature. After all, you can supply your own telephone and use its one-touch or speed-dialing features to call other fax machines quickly.

Document Sending Size

The document width capacity of your fax machine determines the form of information you can send. For example, if you plan to send computer printouts, your fax machine either must have a scanner that can handle documents 11 inches wide or a separate reduction copier.

Scanners that can accept wide documents reduce them to an 8 1/2-inch width before the machine sends them. Documents emerge at the receiving end in standard 8 1/2-inch-width form. Therefore, if you have easy access to a reduction copier—which may reduce the original image more crisply than a fax machine—you may not need the added scanner capability, particularly if you don't foresee sending large quantities of oversized documents.

> **Telephone line noise affects the accuracy and quality of fax messages.**

Take time to evaluate each machine's scanner performance. Some fax machines don't scan the entire width of the page. Like copy machines, they have dead zones at the margins. Be sure to test a machine's scanning ability in the copier mode.

Copier Option

Many people who don't own a separate copier are interested in the fax copier option. The copier option serves two purposes: It lets you preview a fax transmission (showing approximately what it will look like at the other end), and it can serve as a low-volume copier in its own right.

Each use must be evaluated separately. When previewing the copier feature, the more complicated the transmission (such as halftones and other visuals), the less accurate the copier preview of the transmission will be, particularly with fast transmission rates. As mentioned earlier, telephone line noise affects the accuracy and quality of fax messages, and the copier option doesn't account for that. On the other hand, it's useful to have a rough idea beforehand of what the receiver will get, particularly with graphic information.

The quality of a fax machine copy generally isn't as good as one made on a regular copier. Thermal paper, the kind most commonly used by fax machines, has an annoying tendency to turn black under heat; consequently, it doesn't store well. Although fax can double as a photocopy machine in a pinch, if you need photocopying capability regularly, the most prudent course is to buy a photocopy machine.

Resolution

Even the least expensive Group 3 compatible machines support both regular and fine mode resolution. However, if you plan to send detailed graphics or small text, you may require even higher resolution. Unfortunately, the ultra-high resolution modes aren't standard among manufacturers. And you often can send such images only to machines built by the same company.

```
Regular Resolution
Fine Resolution
```

Halftone Transmission

If you plan to send photographs or drawings with tones or screen, you must have the halftone transmission feature. If you'll be sending only black-and-white text and graphics (line drawings), you won't need it. All Group 3 machines can *receive* halftone transmission, regardless of whether they have that feature.

Notes:
Do I need to send photographs or gray scale graphics?
Yes_____ No_____

Superfast Mode

Generally, the faster a message can be sent, the more cost-efficient the fax communication is. Some fax machines come equipped with their own methods for speeding transmission by compressing the data. White-space skipping is an example. When the scanner encounters white space, rather than send several blank lines, it sends the receiving machine a message to skip the entire line—a kind of electronic shorthand.

Unfortunately, if the receiving machine doesn't understand shorthand, you can't communicate in superfast mode. Indeed, the highest speeds can be used only by faxes that use the same shorthand.

> You may find that a machine that can send a page in 20 seconds needs three or more minutes to send a five-page document.

Although different manufacturers may use the same shorthand protocols for high-speed transmissions, machines made by the same manufacturer often enjoy an added advantage of being able to communicate with each other more rapidly.

The way machines determine how to communicate with each other is called *handshaking*. Machines built by the same manufacturer quickly ask each other whether they speak the same high-speed language. Other machines may take longer to find the fastest mutually acceptable code to send and receive information.

High-speed capability can save both time and money. It can be an important feature if you have a large volume of international fax traffic and if you'll be communicating with machines that use the same high-speed shorthand.

Conversely, if your fax communications are mostly local, the extra time you save may not merit the extra cost involved. In that case, the superfast mode would be unnecessary.

Note that actual transmission time may not always meet the theoretical performance specifications. Consequently, if superfast transmission is important, be sure to test the actual capability in a "real-time" situation. Ask the salesperson to send a five-page document to another machine built by the same manufacturer. You may find that a machine that theoretically can send a page in 20 seconds needs three or more minutes to send a five-page document.

Delayed Transmission

By allowing messages to be sent at night, during low-cost calling hours, delayed transmission can cut telephone costs. It also can spread the demand on the fax network more evenly over a 24-hour period.

> **Delaying transmission until international rates are lower can result in substantial savings.**

However, delayed transmission must be programmed, and if only a few fax messages are sent every day or week, it may be more convenient—if slightly more expensive—to send them during regular business hours.

Delayed-transmission capability has the greatest impact when faxing internationally. Often international rates are at their peak during your workday, even though people in another country probably aren't at work for hours. Delaying transmission until international rates are lower can produce substantial savings.

The usefulness of delayed transmission is limited by the capacity of your document feeder. Each document to be faxed must be stored in the document feeder (or the internal memory). Consequently, extensive delayed transmission capability requires a full-featured, high-end fax machine.

Dialing Capabilities

Advanced dialing capabilities (one-touch dialing, speed dialing, alternate-number dialing and stored-number memory) can save time and money.

One-touch dialing is most effective if you have a number of locations you call regularly, perhaps daily or even hourly. Speed dialing, which requires a few extra keystrokes, automates access to a larger universe of fax numbers. Alternate-number dialing allows you to automatically call fax machines that are frequently in use without getting a busy signal. Stored-number memory helps avoid misdialing other fax machines.

Perhaps the only drawback to these advanced dialing features is that they require some programming. As many telephone users learned to their dismay about speed dialing, for example, you may find yourself investing in a feature that no one will use.

Store and Forward

Unless you serve as the middle link in a communication network, store and forward probably isn't necessary. Typically, store and forward is used to take advantage of low local telephone rates. For example, if headquarters in New York faxes new pricing information to the Atlanta regional office, that office then can send the information to local sales personnel.

Store and forward also can be convenient if you receive information from one source that simply must be passed to others. For instance, if you wish to inform several offices of an individual's schedule and you're using a mobile fax machine, you may want to send that information to another fax, which then will forward it to several local points.

Broadcast Capability

Assessing the need for broadcast capability is extremely tricky, because for many companies it will add a new communications ability rather than replace or simplify an existing one.

Broadcasting allows a document to be scanned into a fax machine once and then sent automatically to several preprogrammed telephone numbers. Some machines can preprogram as many as 100 fax numbers for broadcast. Although many fax machines without the broadcast feature can store 100 telephone numbers in their memories, these machines must make multiple copies to send your message to the locations. With broadcasting, the message only has to be scanned once.

How can broadcasting be used? A Chicago manufacturer wants to inform distributors and other sales personnel throughout the country about timely price and product changes. Every Monday morning, updated information is faxed to a list of preprogrammed fax numbers. Salespeople have important new data with which to work, and management knows they received it on time without having incurred the considerable expense of overnight mail services.

> **Polling is useful if you plan to do two-way communication via fax.**

To decide whether you need broadcasting, identify what you do now that can be automated through broadcasting and determine how broadcasting can improve your communications.

Polling

In many ways, polling is the opposite of broadcasting, allowing you to receive information from many locations conveniently. Rather than have each sending machine call the receiving machine—often encountering busy signals—the receiving machine calls and polls the sending machines.

Polling is useful if you plan to do two-way communication via fax. One machine calls another, sends information and then polls the machine to check whether information is waiting for it. This way a single call carries two messages.

Auto Voice/Data Switch

Auto voice/data switch is one of the most complicated and confusing fax features. In theory, an auto voice/data switch should allow a regular telephone user to share a telephone line with a fax machine. If a call comes in for a person, the phone rings. If a fax transmission comes in, the machine automatically accepts it without bothering the person.

Auto voice/data switch is geared for people who wish to receive faxes without operator intervention but whose fax volume cannot justify the monthly charge for an additional phone line. Professionals who spend a large part of their workday talking on the telephone and perhaps have two incoming telephone lines fall into this category, as do field sales personnel and others who work abroad or in different time zones. For example, for a company with a European sales representative who works from home, an auto voice/data switch may avoid an annoying wake-up call in Belgium at 2 AM.

Unfortunately, different manufacturers have different definitions for auto voice/data switch. In one model, for instance, when the switch detects that the transmission is a fax, employees are greeted by the ear-splitting sound of the fax machine responding in machine language. Only seconds later, if no response is received, it switches to voice transmission. By that time the caller has hung up.

For heavy fax users, the first alternative to consider is a telephone line dedicated to fax. With a dedicated line, people calling your phone number will never get the high-pitched whine of a fax machine. Depending on where you live, a dedicated line costs about $300 the first year and $150 a year thereafter.

For light fax users who plan to be present when faxes are received, an auto voice/data switch may be unnecessary as well. The sender can call first to alert you that a fax message is being sent and you then can activate the machine.

Security and Secure Mailboxes

Your need for security depends on three factors: (1) the sensitivity of the information you send and receive; (2) the people who legitimately have access to faxed information on both the sending and receiving ends; and (3) the community of fax users with whom you communicate.

If you use unattended communications techniques routinely with a large number of fax machines and it would be embarrassing or harmful for the wrong information to get into the wrong hands, you should explore the various levels of transmission

Fax messages are like mail without an envelope.

security outlined in Chapter Two. If you wish to send or receive quotes in a competitive bidding situation, you'll want a level of security that at least ensures that the information won't be sent to the wrong machine. If you plan to send confidential internal information within a company, investing in the highest level of security may be prudent.

However, even if the information is sent to the correct machine, fax messages are open and available to anybody standing near the machine at the moment. In essence, fax messages are like mail without an envelope.

That flaw can be eliminated by the use of a *secure mailbox*. When a fax is received, instead of automatically printing, it's stored in the machine's internal memory and is printed only when the correct password is entered. This feature generally works only between machines built by the same manufacturer.

The third security measure is *encryption*. Encrypting data prevents electronic eavesdropping as data move along the telephone lines. The message is scrambled when sent and a key or password is used to unscramble it at the receiving end. Encryption is required only for strict security. Unless you're in the CIA, you probably don't need it!

RS 232 Port

The RS 232 port can be either your connection to the future or a waste of money. In their current configuration, computers and faxes need not be linked, and unless you intend to integrate the two, the RS 232 port is unnecessary.

On the other hand, the new plain paper laser printer fax machines create many opportunities. If you buy a laser printer fax machine and have computers, an RS 232 port enables the fax printer to double as a computer printer.

Even with thermal-paper faxes, however, an RS 232 port might be a useful feature if you're a telex or TWX subscriber. The RS 232 port, along with a special modem and a service called Economy Telex, can drastically reduce your telex monthly bill. (For more information, contact Western Union or your fax representative.)

Many fax salespeople don't understand the purpose of the RS 232 port. This will change as computers and fax continue to combine features. (To find out more about computer/fax integration, see Chapter Five.)

> **Activity reports also can be important if people have uncontrolled access to the fax machine.**

Activity Reports

Activity reports are tools for managing fax usage. In small businesses or among personal fax users, activity reports are probably unnecessary. Telephone charges constitute the primary cost of sending fax, and those costs can be controlled through an analysis of the telephone bill.

However, activity reports are essential for monitoring actual fax flow. Activity reports indicate the time, date and duration of every fax message sent and received. The reports allow a man-

Activity reports are essential for monitoring actual fax flow.

ager to judge whether the fax machine is being used according to company guidelines. In some situations, for example, sending a large document via overnight mail makes more sense than tying up a telephone line and an employee's time for a half hour or more while the fax is being transmitted.

Activity reports also can be important if people have uncontrolled access to the fax machine. For example, through the activity report, one company discovered that one of its engineers was sending faxes to a competitor after hours. Not surprisingly, the engineer hadn't logged those transmissions in the written journal each fax user was expected to maintain.

MODE	CONNECTION TEL	CONNECTION ID	START TIME	USAGE T.	PAGES
RX	886 2 7357956	G-3	08/29 19:07	01'46	01(00)
TX	16037421944	G-3	08/30 07:45	00'45	01(00)
TX	0114414999767:::	G-3	08/30 07:56	00'58	01(00)
TX	16883909	CH PRINTING	08/30 11:19	01'02	01(00)
TX	16883909	CH PRINTING	08/30 12:37	00'49	01(00)
TX	12058776422	G-3	08/30 13:38	01'38	02(00)
RX	45 2 879460	G-3	08/31 06:58	01'20	01(00)
TX	1215561628	G-3	08/31 09:45	00'59	01(00)
TX	17034338339	BANTA H-BURG VA	08/31 10:49	01'35	03(00)
TX	011452879460:::	G-3	08/31 10:56	00'43	01(01)
TX	011452079460:::	G-3	08/31 10:59	00'56	01(00)
RX	18183035770	G-3	08/31 13:12	02'25	03(00)
RX	6814139	G-3	08/31 17:27	01'12	02(00)
RX	021 706 3301	G-3	09/01 07:31	01'23	02(00)
TX	17036840184	G-3	09/01 07:36	04'57	08(00)
TX	0116126814139:::	G-3	09/01 08:00	00'56	01(00)
RX	021 706 3301	G-3	09/01 08:03	01'37	02(00)
RX		G-3	09/01 09:24	03'22	02(00)
TX	16883909	CH PRINTING	09/01 10:36	02'01	01(00)
TX	16037421944	G-3	09/01 10:41	09'56	10(00)

MAJOR PRICE GROUPINGS

People who read fax advertisements know that fax prices are dropping rapidly—so rapidly, in fact, that machines may not fit exactly in the price categories described below. However, fax features probably will continue to be clustered in much the same way, and prices will vary accordingly. With full-featured faxes dipping below the $3,000 mark and basic units costing less than $1,000, selecting the right features is as important as price in choosing a fax machine.

No Frills Fax

The common denominator among fax machines selling for less than $1,000 is that most are only Group 3 compatible and come with very few advanced features. Most print on 98-foot rolls of thermal paper and can receive transmissions unattended. Those two features are the very minimum you should demand in a fax machine.

Beyond this basic configuration, fax machines that cost less than $1,000 vary widely. Many do little more than send and receive. Most do not include document feeders. And although they may come with simple telephone handsets, most do not include auto-dial-ers. Because this group of machines often lacks a display screen, it's impossible to know whether you've called the right number. Not surprisingly, these machines have no activity-reporting features.

> **Inexpensive fax machines are most suitable for light, personal fax usage.**

This class of machine is most suitable for light, personal fax usage and for cost-sensitive buyers. For example, if you periodically work at home and your office has fax capability, a low-end machine might be appropriate for those days when you commute only as far as your study.

However, price-conscious buyers should avoid specials that advertise "Brand Name, New Fax" for $499 or less. Often these machines are old Group 2 faxes, which can take three minutes to transmit a page—as much as 15 times longer than Group 3 faxes. Slow speed alone makes them a bad buy, not to mention their lack of compatibility. The fax revolution really began when Group 3 became available. Don't buy obsolete technology.

Middle-range Fax

Most machines in the $1,000 to $2,000 range come with a liquid crystal display screen, transmit and remote terminal identification and the ability to generate basic activity reports. All communicate at the rate of 9,600 baud and are Group 3 and Group 2 compatible. Some are Group 1 compatible as well. Most have auto-dialing capabilities.

The most distinguishing features of machines in this category are automatic document feeders, polling and internal memory. You should look for a machine with an automatic document feeder that can handle your needs. If you plan to send faxes with multiple pages, the automatic document feeder significantly reduces operator time. Although machines in this price range can poll other machines, they usually lack security features.

These middle-range fax machines can be tricky to buy. In the sales literature, they appear to have all the features of more expensive machines. But because some features are limited, others cannot be used effectively. For example, many machines offer delayed transmission. But if the machine can be programmed to transmit only one document and you send more than one fax each night, you're severely limited. Small automatic document feeders can have the same kind of impact. If you need to use the delayed-transmission feature to send a 20-page document, a 10-page feeder will make that feature useless.

> Mid-range faxes are most suitable for branch offices, small companies and professional practices.

Finally, in addition to the small document feeding capacity, many machines in this range have two serious drawbacks which limit their usefulness for high-volume applications. Most work only with 98-foot rolls of paper and lack automatic document cutting. If you receive large numbers of faxes at an unattended machine, the lack of automatic document cutting may force you to spend more time cutting and collating messages than it took to transmit them. And unattended machines run out of paper often.

Mid-range faxes are most suitable for low- to medium-volume workplaces, such as branch offices, small companies and professional practices, traveling and trade

> With full-featured fax machines you can expect to get every feature currently available.

shows, and personal fax applications. Without a large investment, they provide an excellent way to learn how fax can be used in your specific circumstances.

Full-featured Fax

Full-featured fax machines should be able to serve as the central fax unit in any organization. In the $2,000 to $3,000 price range, you can expect to get every feature currently available. The most useful features are large paper rolls, automatic document cutting, automatic document feeders that can hold 30 to 50 sheets, and internal memory.

In addition, many models in this range have such advanced features as secure polling and secure mailboxes, broadcasting and high-speed transmission. Of course, many of the most advanced features will work only with other machines made by the same manufacturer. So, if you decide you need them, you must be able to control the fax purchases at both the sending and receiving stations.

High-end fax machines soon may come equipped with laser printing. Such models will be able to double as computer printers. They also will offer the advantage of copying on plain paper.

Deluxe Fax

This category will be dominated by equipment that combines several tasks, such as high-speed photocopying or laser printing with fax communication. These machines, which typically cost more than $3,000 per unit, have large internal memories and every feature discussed in this book. In the future, deluxe machines will be Group 4 compatible, which means they'll work on ISDN (digital transmission networks) and will be able to send an error-free page of information to another Group 4 machine in less than five seconds. When Group 4 machines first appear, they'll demonstrate the promise of the future rather than fill the needs of today—so buy with caution.

WHERE TO BUY

As the price of fax technology has dropped, the distribution channels for equipment have multiplied. Like copier machines, fax machines now can be bought directly from salespeople employed by the manufacturer, as well as from retail stores specializing in office equipment. Some high-volume consumer electronics stores, discount operations and even mail-order companies carry fax equipment. In response to increasing consumer demand, department stores also are stocking fax.

> When buying from a manufacturer's representative, you're also buying that company's expertise and ongoing support.

It's important, however, to buy your equipment from a channel of distribution appropriate to your needs. To a large degree, the appropriate channel is dictated by the price range you choose. Discount chains and mail-order companies usually offer the least expensive machines. Manufacturers' representatives offer the most expensive. When buying from a manufacturer's representative, you're also buying that company's expertise and, often, their ongoing support. In some settings, this may be unnecessary. In others, it's vital.

Manufacturers' Sales Representatives

There are several advantages to buying from a manufacturer's sales representative. These salespeople are likely to know and understand their company's various fax configurations. They should be able to spend enough time with you to learn your business, so that they can help to determine which configuration best suits your needs and provide after-sales training to make sure you fully utilize all the features. They should be able to offer a wide range of leasing and financing options. (For more information on leasing, see Appendix A.) Finally, the sales force provides a direct link to the company's service organization. In short, a direct sales organization should be able to take care of you best.

But that service has its price. First, to explore a complete range of offerings, you must talk to representatives of several manufacturers, which can be extremely time-consuming. Second, sales personnel generally ask for the full suggested list price, whereas other channels offer substantial discounts. Finally, because salespeople usually work on commission, they're motivated to sell the most expensive equipment.

Office Equipment Stores

Because office equipment stores generally carry a range of machines from several manufacturers, comparison shopping is easy. And because they're in a retail environment, they may offer competitive price discounts, particularly on older machines. In addition, many offer their own installation and training services.

Nevertheless, office equipment stores have their own drawbacks. First, you're not dealing directly with the manufacturer, and if problems occur, the retailer may not be able to solve them. Second, because the store's salespeople sell a wide range of products, their product knowledge may not be deep. Moreover, salespeople probably won't have time to assess the needs of your business. And retail salespeople may have hidden agendas, steering you toward the units they want to sell rather than those most appropriate for your needs and budget. Hence, if you choose to shop at an office equipment dealer, you must do more homework and understand what you want from your fax network. And you should be sure to check on warranties, service contracts and the location of the nearest repair shop, in case of a malfunction.

For low-volume fax
users, the savings
offered by mail-order
companies may com-
pensate for the lack
of follow-up.

Discount and Mail-order Companies

At discount and mail-order companies, the big benefit is low price.
But you can't expect the salesperson to know anything about the
features of a specific model or the differences between models.
Be prepared to provide for your own setup, installation and
training. However, for low-volume, personal fax users, the
savings may compensate for the lack of follow-up and support,
particularly for people with considerable electronic expertise.

FEATURES ASSESSMENT CHECKLIST

At this point, you should have a good idea of what you want
in fax technology, how much you should expect to pay for a
machine and where to buy it. The Features Assessment Checklist
below can help you evaluate what best suits your needs. Be sure
to take it with you when selecting fax equipment.

Features Assessment Checklist

Compatibility:	Critical	Optional	Unnecessary
Group 3	_____	_____	_____
Group 2	_____	_____	_____

Reasoning:
Number of Group 2 receivers _____
Number of Group 3 receivers _____

Receiving Capacity:	Critical	Optional	Unnecessary
Large	_____	_____	_____
Small	_____	_____	_____

Reasoning:

Average number of faxes received daily _____

Average number of pages per fax _____

Hours machine is unattended _____

Paper capacity needed for unattended operation _____

Doc. Feeder Size: Critical Optional Unnecessary

  Large   _____   _____   _____

  Small   _____   _____   _____

  None   _____   _____   _____

Reasoning:

Plan to use delayed transmission or polling? Yes_____ No_____

Number of pages of average fax message _____

Telephone Handset: Critical Optional Unnecessary

Reasoning:

Is a convenient alternative to integrated handset available?
Yes_____ No _____

Doc. Sending Size: Critical Optional Unnecessary

  Large   _____   _____   _____

  Standard   _____   _____   _____

  Small   _____   _____   _____

Reasoning:

Maximum width of paper to be scanned _____

Number of oversized documents transmitted per month _____

Is reduction copier accessible? Yes_____ No_____

Copier Option: Critical Optional Unnecessary

Reasoning:

Do you want to preview documents before they're sent?
Yes _____ No _____

Would low-level photocopying capability be helpful?
Yes _____ No _____

Resolution:	Critical	Optional	Unnecessary
Fine Mode	_____	_____	_____
Super Fine Mode	_____	_____	_____

Reasoning:
Do you need more than fine mode resolution?
Yes____ No ____
Will you communicate only with fax machines of the same brand? Yes____ No____

Halftone Transmission: Critical   Optional   Unnecessary
Reasoning:
Do you need to send photographs or gray scale graphics?
Yes_____ No_____

Superfast Mode: Critical   Optional   Unnecessary
Reasoning:
Volume of international fax traffic in pages per month ____
Number of fax receivers that support superfast modes _____
Number of fax receivers from same manufacturer _____
Speed of fax transmission in actual time test _____

Delayed Transmission: Critical   Optional   Unnecessary
Reasoning:
  Volume of faxes per month to be sent at low-cost hours ____

Dialing Capabilities:	Critical	Optional	Unnecessary
Auto-dialing	_____	_____	_____
Speed Dialing	_____	_____	_____
Alternate Dialing	_____	_____	_____
Stored Capacity	_____	_____	_____

Reasoning:
Number of fax numbers accessed regularly _____
Frequency of calling regular fax numbers daily or weekly ___
How busy are the numbers, particularly during unattended hours?

Number of telephone numbers that memory can store_____
Who will actually program the numbers? _____

Store and Forward: **Critical**   **Optional**   **Unnecessary**
Reasoning:
Will the information you receive be passed on without comment?
Yes_____No_____

Broadcast Capability: **Critical**   **Optional**   **Unnecessary**
Reasoning:
Current broadcasting tasks_____
Potential broadcasting tasks_____

Polling: **Critical**   **Optional**   **Unnecessary**
Reasoning:
Will you be gathering information from many sources in limited
    time frames? Yes ____ No ____
How often?
Will you conduct two-way fax
    communication? Yes ____ No ____

Auto Voice/Data Switch: **Critical**   **Optional**   **Unnecessary**
Reasoning:
What is your anticipated monthly fax volume? _____
Is a dedicated line cost effective? Yes ____ No ____
Will you be present when faxes are received? Yes ____ No____

Security and Mailboxes: **Critical** **Optional** **Unnecessary**

Security and Mailboxes:	Critical	Optional	Unnecessary
Secure Polling	_____	_____	_____
Secure Mailbox	_____	_____	_____
Encryption	_____	_____	_____

Reasoning:
How sensitive and confidential are the data you plan to send and
receive?
Can you be sure that the intended person is the only one who will

have access to the information? Yes _____ No _____
How closely will you monitor actual fax sending and receiving?

RS 232 Port: Critical   Optional   Unnecessary
Reasoning:
Will you use the fax for telex service? Yes_____ No _____
Is this fax machine an effective printer? Yes _____ No _____
What benefits will you receive today?
What future benefits will you realize from this machine's RS 232 capability?

Activity Reports:	Critical	Optional	Unnecessary
Communication journal	_____	_____	_____
Transmission report	_____	_____	_____

Reasoning:
How many people will have access to the fax?   _____
How damaging can unauthorized fax usage be? _____

MOVING ON

By now you should have a good idea which features—and clusters of features—you need. You also should have an estimate of your price range.

The remaining chapters assume you own a fax machine. They provide information on installing, managing and maintaining this new technology.

4

Up and Running

A FAX MACHINE FRESH OUT OF THE BOX

After you've decided to buy a particular fax machine, how difficult is it to get that machine up and running? In most cases, not too difficult. With a simple machine, you should be able to fax Singapore within an hour after taking the machine out of the box.

Unlike computers, which require days (if not weeks) of training, fax machines can be set up and put to work with a minimum of confusion, provided a few precautions are taken. On the other hand, if you don't work through installation problems carefully at the outset, you may deprive yourself of several advantages.

Key Steps to a Smooth Transition

Getting your equipment to work may be the easiest part of putting fax into operation. The difficult part lies in how you handle the following steps:

1. Decide where to place your fax machine and who will operate it.

2. Choose a long distance telephone company.

3. Have your local telephone company install an additional telephone line.

4. Install your fax equipment and test it immediately.

5. Train your personnel, and encourage fax use by explaining its capabilities.

WHERE TO PUT IT? WHO WILL USE IT?

Your decision about where to install your fax should be determined by how it will fit into your communications network. How do you expect it to be used, and who will operate it? Do you want the fax machine in your own office, in a co-worker's office, in the mailroom, next to the copy machine, close to the reception-ist or all by itself?

Fax Machine Location in Large Offices

In large offices, a single fax machine often is shared by 25 to 50 users. In such situations, it's usually located in the mailroom or in another central location with its own clerical staff. When a message is received, the mail clerk delivers it to the appropriate person.

Although this arrangement prevails, and has been used for telex communications for years, it probably won't be effective in the long run. Unlike telex, the fax machine's operation must be rou-tinely monitored, because most fax machines don't ring bells to signal the arrival of a message. Also, the high cost (and low capacity) of telex limits its use. However, as fax use increases and prices drop, machines will be shared by fewer people receiving more mes-sages.

> In a large company, the mailroom may be the most logical location for the fax machine.

In a large company, the mailroom still may be the most logical location for the fax machine. After all, fax closely resembles mail. But to handle fax effectively, the mailroom must be geared up and staffed properly. That may require extra personnel to deliver fax to the recipient promptly.

Moreover, a company may insist that fax messages be packaged in interoffice mail envelopes or special envelopes designating fax mail to help ensure confidentiality. Designing special envelopes for fax mail can have the added benefit of stimulating fax usage. (Some people theorize that Federal Express became a fixture in office communications in part because of its distinctive envelopes, heralding the urgency of the communications.)

Locating Fax in Smaller Companies

In small and mid-size companies, another logical location for the fax machine is near the central telephone switchboard operator, who often is responsible for sorting incoming mail. Both fax and telephone messages can be distributed the same way, and most employees are already in the habit of checking routinely for telephone messages. If fax messages are placed in distinctive envelopes, recipients passing by the switchboard will see that they have an urgent message waiting for them. Thus, the distribution burden is shared by the recipients and the switchboard staff.

> The more central the fax location, the more you must rely on the clerical staff to distribute the messages.

On the other hand, if you locate the fax machine near a switchboard operator who doubles as receptionist, be sure that the machine is placed where casual visitors can't access it. Remember, faxes don't arrive incognito. An inquisitive guest could read highly confidential information from a fax placed in the reception area.

If you plan to photocopy all incoming fax messages routinely, you may decide to locate your fax next to your copier. The potential advantages and drawbacks of the reception area location also apply to the copier area. First, like the reception area, the copier area is a high-traffic area. Second, the amount of time devoted to handling fax can be reduced by dovetailing it with other tasks.

On the other hand, unauthorized personnel can have access to incoming messages. Furthermore, unless you have personnel regularly stationed in the copy room, someone will have to check periodically for new fax messages—an arrangement that could prove clumsy and unreliable.

Rules of Thumb for Fax Machine Placement

Because fax machines operate almost noiselessly, they can be placed almost anywhere without undue interference with normal activities. Unlike computer printers, fax machines don't have to be encased in a plastic cage to reduce noise. Consequently, the rule of thumb in finding a location for a machine can be its convenience for the people who plan to use it. If the fax machine is poorly located, people will tend to use it less. Your aim should be to maximize appropriate usage.

Because larger businesses have more sophisticated in-house mail systems, it may be beneficial to integrate general fax traffic with procedures for distributing mail. Keep in mind, however, that it may be cost-effective to have one machine per user, particularly if confidentiality is involved.

A mid-range solution is to deploy fax at a departmental or work-group level. One of the incoming telephone lines to that department can be reserved for fax use, and the departmental clerical staff can be responsible for distributing fax messages. Without eliminating the problem of confidentiality, strategic distribution of fax machines throughout an organization often makes more sense than centralization.

Your decision, of course, has important ramifications. The more central the fax location, the more you must rely on clerical staff to operate the machinery and distribute the messages. That means staff must be trained and sometimes redeployed from other responsibilities. The biggest risk, however, is that if the clerical staff doesn't perform well, fax won't be used well and you won't enjoy its full advantages.

> Poor quality transmission forces fax to fall back to a slower rate.

Personalizing fax has drawbacks as well. First, higher-paid employees will waste time fiddling with the machine. Moreover, the greater the number of users, the more messages each user must send to justify the placement, which can result in unnecessary use. Last, if a personal machine is perceived as a status symbol, jealousy and rivalry can develop. Good fax equipment isn't inexpensive. When deciding the location of the fax machine, decide how and by whom you want the machine to be used.

CHOOSING LONG DISTANCE SERVICE

Choosing a long distance service for fax use can be as difficult and confusing as choosing a long distance service for regular telephone use. Of the long distance providers, none yet offers special inducements for fax users.

The first and most obvious question is whether the carrier serves the locations to which you'll be sending faxes. Even some of the large long distance suppliers still have limited international coverage.

Determining the quality of transmissions sent by alternative long distance suppliers is also important. The human ear adapts to and tolerates the quality difference between fiber optic and microwave transmissions. Fax is more sensitive, and poor quality

> **Be suspicious if a salesperson offers you a special telephone line for fax use.**

transmission forces it to fall back to a slower transmission rate.

For example, if lower quality long distance service is 20 percent less expensive per minute, but your faxes must be sent at 4800 baud instead of 9600 baud because of the poor quality of the transmission, you'll wind up spending 60 percent more in telephone costs, instead of saving 20 percent! Furthermore, many fax messages will be garbled and hard to read.

The final element in the selection of a long distance service is availability of special services. Bear in mind that WATS lines can be used as effectively with fax technology as with telephones.

After the AT&T divestiture and deregulation of 1984, most businesses reassessed their long distance suppliers. If you're satisfied with your present supplier, you probably should add the fax equipment to your current network. Otherwise, purchasing fax equipment can provide you with a good opportunity to reassess your long distance options.

After you've decided where you want to place fax and which long distance service you plan to use, you may need to install additional telephone lines. Because new installation often requires several weeks' notice, call the telephone company promptly. When you call, keep the following points in mind:

1. If you can justify the cost, reserve a telephone line dedicated to fax use. Although a fax machine can accept calls through a PBX system, separate lines allow fax machines to answer calls unattended. They also prevent scrambled messages or other difficulties that result when people unwittingly pick up the phone while the fax is in use.

2. You don't need a special line—such as a data line. Fax works just fine over regular telephone lines. Be suspicious if a salesperson offers a special line for fax use.

3. If you're installing a fax in your home office, ask for a residential line, not a business line. A residential line is usually less expensive.

4. If possible, don't have your fax number published in the telephone book. If you do, some people may call directory assistance for

> **Be sure to insist that no features are added to the telephone line you plan to use for fax.**

your telephone number and instead receive your fax number. That can be ear-splitting, as well as frustrating.

5. Think twice about having your fax number listed in any kind of directory. You may find yourself deluged with unwanted fax junk mail.

6. Don't add features to the telephone line you plan to use for fax. Many local Bell operating companies automatically include a package of features for the subscriber to "test" with each new line. Be sure to insist that no features are added, particularly Call Waiting, Automatic Call Forwarding and so forth. Above all, avoid Call Waiting, which signals another incoming call with a loud click on the line. The fax machine doesn't understand the function of this click and misinterprets the information, consequently garbling incoming and outgoing messages.

Fax technology's sensitivity to Call Waiting may present problems for people who want to install fax on an existing phone line also used for voice communication. If you can't dedicate a line to fax, the best solution is to install a second line with a rotor. Incoming calls then are routed from a busy line to the second line. Rotor capability costs only a few dollars per month—not significantly more than Call Waiting—and it's more convenient and not threatening to your fax.

7. Determine the best call area billing system for your fax telephone line. If most of your messages will be local, you may want to invest in extended local area dialing. If most of your faxes will be sent long distance, you probably will want a restricted billing area. Before calling the telephone company, analyze your needs carefully to decide which of its billing options is appropriate for you.

INSTALLING YOUR EQUIPMENT

The procedures to follow for installing and using your machine depend upon where you bought it, the complexity of your system and the number of people you expect will use the system.

The Manufacturer's Representative

If you bought your machine from a manufacturer's sales representative, installation and training should present no problem. In the purchase agreement, you should insist that the representative come to your workplace, install the machine and train your personnel. Because you've paid a higher margin for service, don't waive that advantage, particularly if you've purchased a complex machine with such features as broadcasting and account codes and if you expect many people to have access to the fax.

As a rule, try to have a competent representative explain and demonstrate as much as possible about the machine you buy. Unfortunately, because operating procedures vary markedly from manufacturer to manufacturer, you can't count on other fax users to help resolve problems that may arise.

> If you buy from the manufacturer, you may want to choose a vendor with support activities separate from sales activities.

Not surprisingly, some sales representatives are more interested in making sales than in providing support. So you must determine exactly how the support you've purchased will be delivered. As people who buy new cars ruefully learn, the salesperson who says he or she will be your personal representative to the company is too often unavailable.

Nevertheless, company support and service personnel should be able to install your machine and train your employees better than anyone else. If you make a large unit purchase, try to have an account support representative assigned to you. If you buy direct from the manufacturer (and the extra level of available support is a good reason to go in that direction), you may want to choose a vendor with support activities separate from sales activities.

Office Equipment Dealers

If you buy your fax machine from a retail store, try to elicit as much information as possible about the machine *before* you make the purchase. When you leave the store, you may be on your own. On the other hand, many retailers pride themselves on the product knowledge of their sales personnel. Try to test that knowledge and put it to use.

Before completing the sale, open the box at the store and inspect the machine. Then browse through the instruction manual. The quality of instruction manuals varies widely. Some manuals are virtually unintelligible. Others provide clear and concise guidance.

> If you bought your machine from a mail order company, you decided to invest your own time in exchange for a low price.

Regardless of the quality of the instruction manual, ask the store personnel whether they can program the transmit terminal identification number and the time and date into the memory. If no one at the store can perform these tasks, you may want to shop elsewhere.

In addition to one-time operations (such as programming the transmit terminal identification number), ask a salesperson to review the basic operation as you follow along with the manual. Have him or her program some numbers into the speed-dialing memory and watch you as you program some. Review the paper-loading procedure. Don't leave the store without a roll of paper loaded.

Mail Order or Discount Stores

If you bought your machine from a mail order company or a discount store, you decided to invest your own time in exchange for a low price. With fax technology, that can be a good investment. When the machine arrives at your workplace, unpack the unit carefully and inspect it for damage. Then review the instruction manual, plug the machine in and go through the initial programming procedures for transmit terminal identification, date, time, speed-dialing and so forth.

> Lost faxes can hurt your business even more than lost telephone calls.

Reserve enough time so that when you finish, your machine will be fully operable. Because you basically can plug a fax machine into a phone jack and send a message right away, you should program your features immediately; otherwise, you may never program them. If you've paid for advanced features, it's a waste of money not to use them.

AVOIDING POWER OUTAGE PROBLEMS

Power outage problems pose one of the most serious threats to your fax machine. A power surge or lightening can disable your fax and result in a sizable repair bill. To protect your fax from being damaged, use a power surge protector. A good one will cost less than $50 and provides excellent insurance.

While on the subject of power protection, an Uninterruptable Power Supply (UPS) is a great power protector and allows your fax to keep operating even if the power fails (although you'll need a flashlight to read your incoming faxes). This option need only be considered if you regularly experience power outages and if it's important that you always be able to receive and send faxes.

THE BUCK STOPS HERE

No matter where you buy your equipment, assign responsibility for your fax operations to one person who'll maintain the equipment, train operators and generally stay abreast of the field. Fax technology is improving rapidly. One person in your organization should be responsible for keeping you on the leading edge of communications.

THE FINAL STEPS

After you've installed your fax machine, test everything. Because you may be relying on several different vendors—one for fax hardware, one for telephone-line hardware and one for long-distance services—each probably will point the finger at the other if something doesn't work. All glitches should be resolved before you print your fax number on your business card. Lost faxes can hurt your business even more than lost telephone calls.

Finally, take time to train your personnel—all your personnel—on fax usage. To get the most out of fax, not only must the equipment be running but your employees also must be running to it. Training should consist of a one-time introduction to the new equipment, followed by ongoing suggestions about how to use fax and when not to use it.

MOVING ON

Now that you've installed your fax equipment, all you have to do is sit back and wait for the cost-savings to come pouring in, right? Wrong! In the next chapter, you'll discover how to best manage fax within your company and stay informed about its many changes.

5

Managing Your Fax

In the early 1980s, midlevel employees in many corporations began to discover the potential of personal computing. If management was slow to react, these corporate renegades used personal computers to do their own jobs faster and better, often purchasing machines on their own and using them for a single application, such as word processing or spreadsheet analysis.

Within a few years, many corporations were (and still are) littered with incompatible machines, running dozens of different software packages for the same applications. The power of personal computing ran uncontrolled. Training, service and software costs soared until companies decided to standardize on a single machine. Even if each individual used a personal computer effectively, personal computing often wasn't used effectively throughout the company.

> Without coordination, fax technology can wend its way through a corporation without much control.

Now fax technology has reached a similar stage of development. Individuals and departments are realizing independently the advantages of distributing hard copy instantly. In many companies, the accounting department may want a fax machine for sending reports to and from company headquarters. Then the sales department may begin to use the machine, but worry that accounting is unduly privy to messages and thus want a machine of its own. In large sales departments, a single machine may not suffice. High-volume salespeople may want their own machines.

Without central coordination, fax technology can wend its way through a corporation without much thought, control or care given to its overall use. To maximize its technological advantages, fax must be well used and well managed.

Because every business is different, there isn't a formula for introducing and managing fax use. Nevertheless, in virtually every situation, even in a home office, resolving critical management issues will make fax operation easier on everyone.

Fax technology raises three basic management issues: *information* management, *personnel* management and *cost* management. Information management includes procedures and processes that ensure that fax enhances the flow of information within your company. Personnel management enables the right people to use and maintain fax and the fax network. Cost management ensures that the new technology saves rather than squanders money.

Let's examine these three areas of fax management.

INFORMATION MANAGEMENT

Information management can be divided into four areas: integrating yourself into the fax community; ensuring that the information you *send* reaches its intended destination; ensuring that the information you *receive* reaches its intended destination and is properly stored; and, finally, establishing guidelines defining the appropriate uses of the fax machine.

Entering the Fax Community

Once your fax is up and running, your first task is to inform those with whom you want to communicate that you have a fax machine. This seems obvious, but if you mishandle the way you inform people of your fax capability, you may hinder the development of fax as a routine communications instrument in your setting.

First, decide who should know you have fax. If you bought the machine for your sales department, do you want your vendors to know you have fax? If you bought only one or two machines for your entire operation, are you sure that everybody with whom you do business needs to have instantaneous access to your company via fax? At times, it may be advantageous to have some people send you information by mail rather than by fax.

For example, many public relations agencies now fax their press releases to news agencies and their clients. They believe that fax gives the information an aura of urgency and importance—and it may do just that. However, journalists usually work on a specific story for a specific deadline and cannot react quickly to random press releases. Instead, they generally scan their mail periodically for new ideas. Because they can't be easily integrated into the journalist's work routine, press releases received by fax can be a nuisance.

On the other hand, journalists often need background information on short notice from companies, the government or public relations agencies. For example, a large daily newspaper in California has bureaus more than 100 miles away from the main editorial location. When a reporter requested clippings from the newspaper's library, those clippings used to be sent by courier service, which was slow and inefficient. Now they're sent by fax.

> Before you let the whole world know you have fax, decide if you want the whole world to know.

Thus, although journalists may want information sent via fax in a specific situation, they may not want every public relations agency to know they have fax capability.

Getting the Word Out

Before you let the whole world know you have fax, decide if you want the whole world to know. Of course, once you've determined the community of people with whom you want to communicate via fax, you have to let

> **By restricting fax use to its intended purposes, you can control its growth within your organization.**

them know your fax number. The easiest, most efficient and comprehensive way to inform them is to send an announcement via the fax machine. The announcement can be short and to the point: "ABC Company is pleased to announce that its new facsimile number is 888-444-4444. Please add this listing to your internal fax directory and route this information to the proper departments. Our regular mailing address is 5 Main Street, Centertown, NC 58888, and our voice telephone number is (888)-444-8888." The inclusion of your mailing address and telephone number allows recipients to photocopy all the information and update their telephone directories.

Of course, a simple message via fax probably won't be sufficient to inform everybody that you have fax. Add your fax number to the business cards and stationery of the people who are responsible for authorizing fax usage.

Once again, the fax number should be restricted to those who'll use the fax network. By restricting fax use to its intended purposes, you can control the growth of fax within your organization and force employees to justify their demand to be included in the network.

Increasing Your Fax Audience

Once you've identified your fax communication audience, try to draw them into your network as many ways as possible. In fax technology, it's better to attract too many people than too few. Because fax technology is so effective, it quickly can become overused and overburdened. And although you want to restrict fax usage to appropriate purposes, controlling it too severely can prevent the maximum return on your investment. If usage gets too heavy, more machines can be added to

> **Because the transmittal sheet is the most important tool to help get your message to its recipient, no fax message should be sent without one.**

your network. But if fax usage isn't heavy enough, you can never measure the business and efficiencies you may be missing. When in doubt, you'll do better by including rather than excluding potential users.

Checking Your Transmissions

Fax messages are like mail without envelopes, and you may have no idea how a fax message is handled at the receiving end. Often your fax message passes through several hands before it reaches the person for whom it was intended.

Always include a complete transmittal sheet with every fax message. Because the transmittal sheet is the most important tool to help get your message to its recipient, no fax message should be sent without one. The standardized transmittal sheet provides the following information:

1. Identity of the sender.
2. Identity of the recipient.
3. Length of the message.
4. Person to be contacted if there has been a transmission problem.
5. Additional instructions to the receiving operator, such as alternate recipients, cc:'s and so forth.

The first element in the transmittal sheet should include your company name, address, telephone number and fax number. Then the receiving party can log the receipt of the information. Many companies use their letterhead for the transmittal sheet. If you choose to do that, it's prudent to include the company identification a second time in larger type, because faxed versions of letterhead stationery are sometimes hard to read.

The second element of the transmittal sheet often resembles a memo. It should provide spaces both for your name as sender and for the recipient's name. Leave several spaces for the recipient's name if you want copies of the fax message sent to several people. The fax operator who receives it can copy and circulate the information.

FAX

Date: _____

TECO U.S.
Chapel Hill, NC 27515-2496
Telephone (939) 732-9449
Telefax 939 142-3898

Attention: _____

Company: _____

From: _____

Number of pages_____including this page

Comments:
_____
_____
_____
_____
_____
_____

Next, you should note the number of pages, including the transmittal sheet, that have been sent. For many reasons—usually telephone noise—one page in a long fax document may be lost in transmission. Moreover, the communication can be garbled or otherwise rendered unreadable. This information allows the receiver to make sure the complete document has arrived.

Don't forget that the page numbers in fax transmissions don't necessarily conform to the page numbers of the document you're sending. For example, you may be sending pages 10 to 25 of a report, so always number the actual pages of the fax transmission. Furthermore, if you send your fax in sequence—that is, from page one to nine—page nine will be the top page in the pile at the receiving end, and the transmittal sheet will be at the bottom. By including the page numbers on the transmittal sheet, you let the receiver know immediately if there's a problem.

Because fax information generally is urgent—or at least timely—you may want to list several alternative recipients for the receiving operator to notify. If only one person is listed as the recipient of your fax message and that person is unavailable, the delay in reaching the recipient may cause your information to lose its timeliness and effectiveness.

The last area should be large enough to include short notes and information. In some cases, the whole fax message may fit into the comments area. In others, the area can be reserved for special handling instructions, such as "Notify Jim at the Dallas office when this arrives," or "Immediate response requested."

The purpose of the transmittal sheet is to allow anyone who receives the fax to make sure the entire message has arrived safely, collate it and forward it to the proper destination.

Backing up Your Fax

At most companies, mail-handling routines are more firmly established than fax-handling routines. Because you often can't be sure your recipient has received your fax, you can use one of several back-up methods.

For example, you can mail routinely a copy of a fax message after you've faxed it. Although it will arrive several days later, the letter often will be welcome. If your recipient has been out of the office for several days, the fax message may be lost or misplaced, but the mail will get through. Also, mail is easier to file than fax transmissions, in which cutting lengths are inconsistent and thermal paper is more difficult to handle than bond paper.

A second strategy is to notify the intended receiver by telephone that a fax has been, or will be, sent. Though it may seem redundant to call and then fax a message, you'll often save time by reducing errors. In one company, for example, an executive found himself talking to an unusually demanding customer for hours on the telephone, only to find the information was misunderstood because the fax clarifying the information hadn't been received. Now the customer calls to alert the executive that he'll be sending questions via fax. The executive finds the answers, calls and then faxes them back.

> **Instead of being redundant, fax and telephones often work best in tandem.**

Similarly, a manager may call a vendor to request certain product information. Rather than sit at the phone waiting for the salesperson to fumble through data sheets, the manager can instruct the salesperson to fax the information and call as soon as the information has been received. Instead of being redundant, fax and telephones often work best in tandem.

The Right Tool for the Right Job

Because fax transmissions often are seen by more than one person, you must be extremely cautious about the information you send. Some information can be damaging or have unintended consequences.

For example, one small company lost a bill from a supplier—an honest mistake. In this case, however, the supplier didn't send an overdue notice. Instead, an overeager accounts receivable clerk faxed a message threatening to cut off service if the bill wasn't paid in five days. A clerk at the receiving end interpreted the message to mean his company didn't have enough money to meet its obligations, including the payroll. Morale among the employees was shaken until the owner resolved the problem. But she was so angry at the indiscreet and aggressive way the overdue bill was handled that she immediately began to search for a different vendor for that service. In the wrong hands, even innocuous information can seem ominous.

If the information sent is indeed ominous, a faxed message can precipitate dramatic and undesired response. For example, a Los Angeles bank executive notified branch managers that the commission structure for the loan representatives would be reduced in the wake of the bank's poor quarterly performance. His intention was to have the managers inform the loan officers. However, for the sake of speed and efficiency, he sent the notice via the fax network. When they prematurely learned about the changes, more than 60 loan representatives quit, decimating the sales staff. This kind of information would have been handled best in face-to-face meetings or by telephone.

> If you wish to fax personnel records and payroll data, you may need a fax machine located in a secure area.

Protecting Confidential Information

Confidential information must be safeguarded. The fact that you have a top-of-the-line fax machine in a central location doesn't mean that you have to channel all fax traffic through it. If you wish to fax personnel records and payroll data, for example, you may need a small fax machine located in a secure area. An alternative to a second machine is a fax machine with extensive internal memory and secure mailboxes (see Chapter Two).

Checking Your Incoming Messages

Check for fax messages regularly. For fax to work well, a mes-

Because the technology is new, take time to establish routines to ensure that fax mail isn't lost.

sage must be received promptly by the person for whom it's intended, acted upon and, when appropriate, filed. In most situations, fax messages will be received by a fax operator instead of the ultimate recipient.

Fax operators who have other designated responsibilities should be instructed to check the fax machine on a regular schedule. The advantage of fax is its speed. If the machine is checked for messages only once or twice a day, that advantage is lost. Moreover, if incoming messages pile up, they will be harder to sort and collate. Because the technology is new, take time to establish routines to ensure that fax mail isn't lost either before or after it gets to the receiver.

Routing and Storing Fax Messages

The correct procedures for routing and storing incoming fax traffic depend both on who your fax operator is and how you routinely handle other forms of incoming information, particularly mail. For example, many executives have instructed their secretaries to copy every significant incoming letter and store the copy in a correspondence log as a safeguard if the original is lost or misfiled.

If you use that system, follow it for fax messages as well. If you have other methods in place to safeguard information, you'll want to integrate fax messages into those systems. If you don't have a system, you should establish one. Fax messages—particularly those on thermal paper—simply aren't as sturdy and easy to handle as information on regular bond paper.

Protecting Your Fax Documents

In most cases, when a message is received, the operator collates it, attaches the incoming transmittal sheet and enters it in a log. On the basis of their experience with telex, some companies copy every fax message that arrives and file the copy in a special location. Consequently, if a fax is lost, the information can be retrieved by using the log to determine which day it was received.

Some managers object to this approach on the grounds that a central file that intermingles fax messages with messages for others violates the confidentiality of the correspondence. In some ways they're right. Central copying and filing usually infringe on privacy. A solution could be to automatically copy the fax when it arrives and then deliver that copy with the original for filing purposes.

If you choose not to adopt a policy of having the fax operator copy messages as they're received, you should instruct the designated recipient of a message to copy it immediately, especially if the information is to be kept for any length of time. Thermal paper isn't as sturdy as bond paper, doesn't last as long and is easily destroyed when exposed to heat.

Streamlining Fax Distribution

Once you've established procedures for removing faxes from your machine promptly and safeguarding the information you've received, you need to be sure that faxes are delivered promptly or picked up regularly by the intended recipient. Senders expect their fax messages to get through, and should be told if those expectations can't be met.

> **Devise ways to alert the sender if the recipient hasn't received the message.**

There are several appropriate methods for ensuring prompt delivery. Fax messages can be delivered by the fax operator (in which case, special fax envelopes should be used); you can install special fax mailboxes; messages can be placed together with telephone messages (this won't work if you use a voice messaging system); you can buy a fax machine for every desk that receives fax messages, an expensive alternative.

Obviously, the option you choose reflects the way you do business and the fax network you've established. No matter how you distribute fax messages, you should establish clear guidelines and policies on how to respond to incoming fax messages if the intended recipient is absent from the office and how to alert people on the road that they've received a fax message. Senders know that they've successfully sent a fax transmission, but can only assume that it has reached the intended recipient.

Devise ways to alert the sender if the recipient hasn't received the message. The most direct way is to send a return fax stating that the recipient isn't in the office and won't be available for a specified length of time. Of course, this requires either that recipients inform the fax operator of their schedule or that their support staff handle the communication.

Fax Junk Mail

Not long after a small manufacturing company bought its first fax machine, the firm received an intriguing message via fax. A company offered to send them a $20 ice chest in exchange for the numbers of 40 additional fax machines. The company said it wanted to fax promotional material about low-cost fax paper to the 40 fax users.

> Before you list your number in a fax directory, consider the consequences.

This incident illustrates a major trend and nuisance for fax users— fax junk mail. Because they work over telephone lines, fax machines are susceptible to the solicitations of charities and junk calls. Once marketers have your fax number, you can do little to prevent their sending information, particularly if you can't evoke security measures because your fax machines must be available for transmissions from prospective customers.

One sure-fire way to receive a load of fax junk mail is to list your number in a fax directory. The logic behind fax directories seems as compelling as the logic behind telephone directories. If somebody you don't know wants to reach you via fax, a directory is the easiest way to find your number. Or, as one directory advertises, "With a directory, your fax capabilities have no limit."

But, unlike telephones, fax machines aren't general-purpose communication instruments. For now, fax enjoys a degree of urgency. It often demands more attention than even overnight mail, and you don't want to have to sort through a lot of junk fax to find your message. Indeed, if your machine is jammed with unwanted fax messages, important messages easily can be lost.

Safeguard your fax number. Don't expose it thoughtlessly. Before you list your number in a fax directory, consider the consequences.

Just as you protect your own fax number, protect the fax numbers of others. By responding to a junk fax solicitation requesting a copy of your communication register, you're giving away dozens of other people's fax numbers.

If you want to stop fax solicitations, call the company sending them and ask to be removed from its list. You may want to prepare a form requesting that your number be taken off a sales list. You then can fax it to the soliciting party. Reasonable companies will comply with your requests. After all, they want your business, not your wrath.

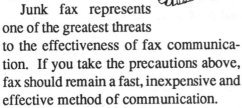

Junk fax represents one of the greatest threats to the effectiveness of fax communication. If you take the precautions above, fax should remain a fast, inexpensive and effective method of communication.

If junk fax does get out of control, changes in fax equipment will be required to screen incoming fax communications. One possible method might be to require pre-registration of all fax machines. Machines would have an allowed access list that would be programmed with the fax numbers of all machines authorized to send messages to it. If a machine not on the list tried to send to your machine, it would be refused.

With appropriate software on PC/FAX, another method might be to allow one page of a fax to be transmitted from any machine. That page would be a request for registration. If the receiving machine operator didn't then program the machine's identification information into the access list, no further faxes would be accepted.

Internal Fax Abuse

Outside abuse in the form of junk mail is hard to counteract. Inside abuse in the form of employees using the fax machine when they shouldn't can be curbed by guidelines specifying the type of information that should be sent by fax—guidelines similar to those governing the use of overnight mail. The rule of thumb is that messages should be faxed when they're urgent, when the quality of the paper they arrive on isn't critical and when they're short enough to be easily transmitted by the fax machine.

Because fax transmission is generally inexpensive, the guidelines need not be as stringent as those for overnight mail. Nevertheless, if a machine is to be used by the sales department to receive inquiries and send quotations, you don't want its time dominated by the central administration office checking supplies at branch offices. If it is, you probably should buy another fax machine for your administrative staff.

Use the fax machine to communicate only specific types of information. After you've bought a fax machine, the uses for it are likely to proliferate. Take time to develop procedures to ensure that you're integrated into your fax community, that the information you send and receive reaches its destination and is properly stored, and most importantly, that fax use is restricted to its intended purposes even as its usefulness to others grows.

Legal Considerations

One of the most commonly asked questions concerning fax is, "Is it legally binding?" That is, is a document that is signed and transmitted via fax a legally binding agreement? The simple answer is yes. But like any legal question, it's not quite so simple.

There are two basic requirements. The agreement must be in writing, and it must be delivered. Agreements that are telexed to another party have been found to be valid; but, unlike telexes, faxes include a signature.

While the agreement may be legal and binding, it's best to have the original signed document sent to you. If you ever have to go to court, the "rules of evidence" require the best available evidence, which in this case would be the original document. You can subpoena the original, and if it can't be produced, the fax then can be used as evidence. While no specific challenges to faxes have been brought to a court at this time, xerographic copies of agreements have been upheld repeatedly when the original isn't available.

PERSONNEL MANAGEMENT

Although fax machines can work unattended, you must attend to people using the fax. The way people use the machine determines its efficiency. Personnel issues usually involve two groups of people—those who operate the machines and those who manage the fax network and fax usage.

Do You Need an Operator?

Fax transmissions are like weddings. You spend far more time preparing for them—preparing documents and transmittal sheets, looking up the appropriate fax number—and more time cleaning up afterwards—receiving the fax, copying and distributing the message, filing the information—than it takes to send that information over the telephone line. Someone has to take all those steps; the question is, who?

> In general, fax operation is a clerical task.

In general, fax operation is a clerical task. Consequently, the responsibilities for sending and receiving fax should fall to the clerical staff, either individual secretaries and clerks or a centralized staff. A centralized staff is perhaps more efficient for receiving fax messages. One person can monitor the machine or machines routinely for incoming messages, copy those messages and then forward them to the recipient.

Incoming fax is like mail, particularly in companies with a limited number of machines. One person or a small staff can handle all the incoming material and distribute it.

Outgoing fax, however, is like the telephone. People want to send their messages at a moment's notice. They don't want to be dependent on someone else's work routine or have their priorities reordered by someone else's priorities. Consequently, sending fax probably should be a distributed task, and several people in the organization should be trained to send fax. For outgoing messages, you probably won't want the fax machine to become the "turf" of one individual who's the only person trained to use the machine.

Although fax transmission is primarily a clerical task, all personnel who will create fax messages should learn how to send them as well. And while you don't want managers and executives to waste their time operating the fax machine during the day when support staff is available, you may want them to be able to send faxes if they work at night or on weekends.

In essence, a two-step system seems best for fax transmission—one that reflects the dual nature of fax. One or a select group of people should be responsible for monitoring the inflow of fax messages and logging and safeguarding the information.

Sending responsibilities, however, should percolate throughout the organization. The one exception to this system is delayed transmission. Because most fax machines have fairly limited delayed-transmission capabilities, one person should be charged with queuing up the messages to be sent each night. To facilitate the control of delayed transmission, you probably will want to set up a file near the fax machine for messages scheduled for delayed transmission.

Company Fax Directories

One further issue concerns the establishment of a centralized internal fax directory, listing every number to which a fax is sent. On the one hand, such a directory institutionalizes your fax community and prevents the loss of fax numbers due to personnel changes. On the other hand, many people consider the fax numbers confidential the same way they consider telephone directories confidential and personal. They want to control access to the numbers they've accumulated. The only way to resolve this problem is to meet with the people who communicate via fax and make a joint decision. An internal directory, even if incomplete, is essential.

Encouraging Fax Use

After the fax machine has been successfully installed, the fax manager should establish a training schedule for all appropriate personnel.

When people start to use fax, the fax manager should be on the lookout for innovative uses of the technology and should publicize them in the company newsletter and by word of mouth. This publicity serves three purposes. It reminds people who're still doing business the old way that they should consider using fax. It allows fax use to grow as part of your communications network. And, the more innovative you become with fax, the better the return on your investment.

Keeping the Machines Fit

The fax manager must be sure that the fax machine remains up and running. This requires careful evaluation of the need for a service contract. With few moving parts, fax machines are generally reliable pieces of equipment; but some models, such as laser-printer-based machines, require periodic service.

Staying Up to Date

The core elements of fax technology probably won't change dramatically over the next several years. Nevertheless, as demand for fax grows and it becomes more of a mass-market item, you'll be able to get more features and lower prices. You need a fax manager to monitor continuously the possibilities and opportunities opened up by new generations of fax machines.

Some key developments you can expect in the next few years include an increase in memory, which will allow greater use of delayed transmission; widespread use of bond paper instead of thermal paper, which will provide better quality faxes and may eliminate the need to copy every fax; and further integration of computers with fax machines, which will give faxes more intelligence and flexibility.

One person should be responsible for analyzing those changes
and deciding when to upgrade the
fax network. Because fax can pay
for itself quickly, your company

> Once your fax is in full
> operation, you should
> see a dramatic decrease
> in the amount of over-
> night mail you send.

may benefit from upgrading regularly so that you always have the
most appropriate, cost-effective technology.

COST MANAGEMENT

The fax manager's final responsibility should be cost control.
Fortunately, calculating fax costs is a fairly straightforward procedure. The main tool is the activity log and communication
journal, which details where each fax was sent and how long it
took to send it. The log can be collated with the telephone bill to
determine the telephone charges associated with fax.

The other costs are supplies, equipment and operator time. In
small operations, it's often most convenient to include fax costs
within the larger categories into which they fall—telephone
costs, office supplies and personnel.

Once your fax is in full operation, you should see a dramatic
decrease in the amount of overnight mail you send. If your
overnight mail budget remains the same or increases, your
company either isn't using fax efficiently or is growing rapidly.

SOME RULES FOR MANAGING FAX: A SUMMARY

1. Before you let the whole world know you have fax, decide
whether you want the whole world to know.

2. Restricting fax use forces employees to justify their demand to be included in the network.

3. When you've identified your audience for fax communication, try to draw them into your network through as many means
as possible.

4. Use the fax machine to communicate only time-sensitive
information.

5. Provide a transmittal sheet with every transmission.

6. Check for fax messages regularly.

7. Senders expect their fax messages to get through. If those expectations can't be met, alert them.

8. If your faxed information is confidential, act accordingly.

9. Use back-up forms of communication when appropriate.

10. If you plan to keep fax messages on file, copy them onto better paper.

11. Safeguard your fax number against fax junk mail. Don't expose it thoughtlessly. Don't list yourself in a fax directory without considering the consequences.

12. Always sign your faxes. As with letters, a signature makes the fax more personal and more compelling.

13. If you can't have your fax machine on 24-hour alert for incoming messages, be sure to specify on your transmittal sheet the hours when you can receive fax.

14. Request fax numbers from those with whom you do business.

15. Don't demand instant response from the receiver unless you've agreed to quick turnaround beforehand.

16. Keep fax numbers accessible to all employees.

17. Experiment. The payoffs can be dramatic.

6

PC/FAX:
Friend or Foe?

The integration of fax with personal computing is the wave of the future in communications. However, while current technology combines the two, it's important to weigh the advantages and disadvantages before you add fax capability to personal computers.

If you don't have at least a rudimentary understanding of computer jargon; if such terms as megabytes, incompatibility, ASCII, half slot boards and application programs mean nothing to you; if *.pic

> Merging data processing and telecommunications into a completely integrated system will be a major trend in office automation.

and *.tif look like typographical errors, this chapter may make little sense. To recognize the potential and pitfalls of PC/FAX, you must have a basic understanding of personal computing.

COMBINING FAX WITH PERSONAL COMPUTERS

Over the next ten years, merging data processing and telecommunications into a completely integrated system will be a major trend in office automation. There's no doubt that the computer is the foundation of the automated office. People not only will continue to generate more and more documents via PCs, but they also will want to transmit that information to and from their PCs.

Until recently, fax machines and personal computers had only a paper connection. Information produced with a personal computer had to be printed, taken to the fax machine, scanned and transmitted. At the other end, the information was received by a fax machine. If receivers wished to store the information in their own computers, it had to be re-entered through the keyboard. The system was slow, tedious and inconvenient, particularly for people who generated complex or lengthy documents.

ELECTRONIC MAIL VS. PC/FAX

Electronic mail—currently the major medium for computer communication—can be easy once you master it, but learning how to telecommunicate is not a simple task. The communications protocols in both the sending and receiving machines must match, and time delays built into the different telephone systems around the world follow no particular pattern. Therefore, an important transmission from Hong Kong might never get through, because the sending modem could hang up if the Hong Kong telephone system paused for a moment.

Setting up an effective telecommunications network often requires lengthy, expensive, individual consultation. Even a clear understanding of the details of telecommunication doesn't ensure that messages will be transmitted accurately if there's line noise.

Telecommunicating graphics information is even more complex. Graphics file formats often are incompatible, and not all communications software supports all graphics formats. Indeed, to send graphics information by electronic mail, software used by both sender and receiver not only should be compatible but identical. Unfortunately, that isn't usually the practice. As a consequence, even people who use electronic mail (E-mail) routinely for written documents generally send graphics information by overnight mail.

Because fax sends all information in a standard graphics format, it was inevitable that fax technology be integrated with computer technology not only to compete with E-mail but to allow instantaneous communication of computer-generated graphics. As an added benefit, PC/FAX conveniently links personal computer users with the expanding fax community.

With a burgeoning market, it's not surprising that more than thirty manufacturers now produce PC/FAX technology.

UNDERSTANDING PC/FAX

Although some companies offer PC/FAX in the form of an external box to be plugged into a PC's serial port, PC/FAX is most often an add-on board that is inserted into one of the computer's internal slots. With a board installed, a user can convert computer-generated documents into Huffman code—the code understood by fax machines—and send that information over regular telephone lines to any other fax machine in the world. In turn, it can receive information from any fax machine. Invariably, PC/FAX machines are Group III-compatible. A small number also are Group II-compatible.

An add-on board enables the computer only to send and receive fax messages. In many ways, a PC/FAX board is like a specialized modem, converting electronic data into a form that can be transmitted. However, to have your computer fully emulate fax machines, you must add a scanner to input externally generated documents. Your existing printer can be used to print incoming fax messages.

Fax Boards

A fax board generally takes either a whole or half slot, and a scanner may require a second slot. Because slots in PCs are at such a premium, some manufacturers offer fax boards integrated with both a high-speed modem and the scanner interface. At least one board includes a small computer system interface (SCSI) as well, so that other peripherals such as a CD-ROM or laser disk drive can be added. Though they may cost a little more, integrated PC/FAX boards are slot-savers.

Optical Character Recognition (OCR) vs. Scanners

A scanner reads a document—text or graphics—dot by dot, makes a bit-mapped image and stores it as if it were a graphic. The information, therefore, can be manipulated only through graphics programs, provided your fax software can convert Huffman code into a compatible computer graphics format. OCR readers start with a scanner, add either a dedicated computer and software, or software only to translate the text images from the document into text codes understood by the computer.

> Although expensive, OCR systems are more than 99 percent accurate; low-end systems are less reliable.

The OCR software looks at each character in the document and compares it to known characters. When it finds a match, it assigns the appropriate code, called ASCII, to the character.

Consequently, text which is read into the computer via an OCR reader can be manipulated with such computer software as word processors. To date, the problem with inexpensive OCRs has been quality. Because most are only 95 percent accurate, a document of 250 words may have 12 or more errors on it. Obviously that would require tedious editing for all documents read into the computer via OCR. Although expensive ($20,000-plus), OCR systems are more than 99 percent accurate; low-end systems are less reliable. Several companies sell add-on OCR software that can be used with many scanners, but that software is new and unproven.

Manipulating Fax Information

Most fax software packages include utilities that convert a fax image to other graphics file formats. After the conversion, the fax message can be changed by using a compatible PC paint program. Most boards minimally support PC Paintbrush and Dr. Halo.

However, the fax message is understood by the computer as a graphic—not as characters. For example, you can't receive a spreadsheet via fax and simply

> **When you turn your computer into a fax machine, you retain all the intelligence of the computer.**

import the numbers to another spreadsheet for further analysis. As the new OCR add-on software discussed above matures, this limitation also should disappear.

Fax Board Prices

Depending on their features and specialized capabilities, fax boards range in price from $395 to about $1,495. If you already have a printer, be prepared to spend about $1,800 or more to add fax technology. On the other hand, when you turn your computer into a fax machine, you retain all the intelligence of the computer. Thus, for $1,800 or so, you get all the features that you'd find in the most advanced fax technology, including a large amount of memory and complete programmability. The same functions in a stand-alone fax machine could cost as much as $3,000. And the price is dropping. At the time of this writing, a combination fax board and scanner was available for $1,000.

COMPATIBILITY ISSUES

Until recently, PC/FAX was confined to the IBM-compatible world. However, fax for Macintoshes is just beginning to appear in the market. Of course, once a board is installed, all Group 3 fax machines can communicate with each other, whether they're free-standing or reside in an IBM PC or a Macintosh. In other words, a fax in a personal computer can communicate with all other Group 3 faxes anywhere in the world.

Nevertheless, several compatibility issues remain with respect to hardware. First, the scanner must be compatible with the fax board itself. Many fax boards support only their manufacturers' scanners. Second, because fax information is graphic, it must be compatible with the computer's graphics board and monitor. Some people argue that CGA-level graphics are inadequate for heavy fax use and that EGA should be used.

To view received fax documents, you need good display resolution. Resolution on computers is expressed as the number of horizontal and vertical dots that can be displayed on the screen. Higher resolution graphics cards and displays make viewing faxes easier and faster by showing more of the document at one time on the screen. Hercules, EGA or VGA graphics are recommended for this use. Finally, the fax board must be compatible with the output device, generally the printer. Most manufacturers support Hewlett-Packard, Epson and PostScript protocols.

Ironically, you may run into some problems if you use a laser printer to output fax messages. Laser printers print at 300 by 300 dots per inch. Fax messages are sent at either 98 by 200 or 196 by 200 dots per inch. When you try to convert between resolutions that aren't even multiples, the image can be somewhat distorted. If you

> **Storage space becomes even more valuable if you plan to save your fax messages on disk.**

don't convert (and many boards aren't equipped with conversion software), a square sent in standard fax resolution will print as a rectangle.

Moreover, when you receive a message, you get only the type styles supported by the sending machine. This means that even if you've invested heavily in different fonts for your laser printer, they cannot be used to receive fax messages.

Other Hardware Considerations

Graphic images require large amounts of disk space. To convert a standard 8 by 11 1/2 inch image into 200 dots per inch requires nearly half a megabyte of storage. Even after the data are compressed, it can still take up 50K in storage space. Therefore, you need to reserve at least three megabytes of space on your hard disk for PC/FAX.

Storage space becomes even more valuable if you plan to save your fax messages on disk. Because some fax boards cannot store a compressed version of the message, you may need a huge hard disk to store just a few messages. Because even compressed messages use 50K per page, you won't be able to store many messages on disk.

In essence, this means that PC/FAX can be used efficiently only to capture and print messages. Most users must routinely go to hard copy if they don't want their hard disks overloaded with old fax material.

RECEIVING MESSAGES VIA PC/FAX

Although they've been on the market a relatively short time, a third generation of PC/FAX boards already has been developed. In the first generation, PC/FAX ran in the foreground. In order to send or receive a fax, you couldn't do anything else with the computer. It had to be "dedicated" to fax usage.

In the second generation of PC/FAX, fax operations took place in the background but still called on the computer's central processing unit (CPU). That is, a user could continue to work on a specific application, but the computer's performance could be degraded as the CPU was called on to perform fax tasks. Furthermore, second-generation boards sometimes interfered with RAM-resident programs.

The latest generation of fax boards come with their own CPU, usually an 80186 chip, and 256K RAM. As a result, they work entirely in the background and never call on the computer's CPU.

Consequently, the process of sending and receiving faxes can be completely invisible unless the user is in the middle of an application that calls heavily on the hard disk itself. In such a situation, some minor delays could occur. When a fax is being sent, the user is signaled by a set of red LED lights attached to the side of the monitor or another convenient place, much the same way that lights signal modem or disk activity.

Although background operation eliminates the need to dedicate a personal computer to fax use, a user must still go into the fax program and check the activity log to ensure that a message has been sent successfully or to retrieve a recently received message.

INSTALLING PC/FAX

If you've ever taken the hood off your personal computer, installation should take 15 minutes or less. Most boards come with their own diagnostics. Some even have instructions detailing what to do if you don't understand the questions that are being asked in the course of the installation routines.

LEARNING TO USE PC/FAX

For some boards, sending fax is a two-step process—point to a file, point to a telephone number and the fax is on its way. Interestingly, advanced fax features such as polling, broadcasting and auto-dialing seem easier to master and more convenient to use on a personal computer than on a stand-alone fax machine. And they're much easier to learn than advanced modem communications techniques.

Differences in Fax Board Operations

Different fax boards convert ASCII and graphics formats into fax codes in different ways. Some boards convert the information while the fax message is being sent or received. Others complete the conversion off-line. Although off-line conversion takes longer to send a fax message, multiple transmissions of the same message can be completed more quickly.

Still other boards take a middle approach. They begin the calling and the conversion process at the same time but convert each page before they send it, filling an on-board buffer with the next page of the message. As a result, the first page of the transmission moves slowly, but the pace for the rest of the pages picks up.

> Material generated on a personal computer and transmitted via PC/FAX often looks better at the receiving end than material sent by stand-alone fax.

Another major difference is the ability to preview what the fax message looks like in fax format.

Management Reporting Features

Most new-generation PC/FAX boards come with built-in communications journal and activity log reporting functions. Typically, the activity log and communications journal automatically record the file that was sent, the recipient (perhaps only by the telephone number), the number of pages involved and the time it was sent. Unlike stand-alone fax machines, some fax boards omit the length of time the transmission required.

Hard Copy Reproduction

Interestingly, material generated on a personal computer and transmitted via PC/FAX often looks better at the receiving end than material sent by stand-alone fax. No matter how good a fax's scanner, there always will be some degradation in the image as it's scanned into the machine; and because PC/FAX generates the image internally, the scanning process is eliminated and the image is sharper.

For this reason, with PC/FAX, other fax machines can sometimes function as remote, high-quality printers. Suppose, for example, you have several offices with fax machines and computer equipment but your standard computer configuration has only nine-wire dot matrix printers or high-speed draft-mode printers. If a message is generated on computer and sent via PC/FAX to a remote facsimile machine, it can be printed at nearly 200 by 200 dots per inch. In some applications this may represent a huge leap in quality.

Recognizing that stand-alone fax machines can serve as remote printers for PC/FAX transmissions, some companies have devised boards that transmit AutoCAD and Hewlett Packard Graphics Language (HPGL) drawings automatically via PC/FAX. An engineer who creates a drawing using AutoCAD and HPGL can send it to a remote location without ever printing it at the work site. The program automatically will cut the file into strips that can be reassembled to make a full-sized drawing printed at nearly 200 by 200 dots per inch at the receiving end. This reduces time substantially and provides significant convenience. Moreover, the drawing at the other end is higher quality than if it had been scanned into a facsimile machine in the first place.

> **Unfortunately, PC/FAX boards don't respect the page makeup protocols in computer software.**

CAD/FAX, as it's called, works only when both machines are equipped with compatible boards. It's the first in a series of specialized applications that fax board manufacturers plan to develop and market over the next several years.

Unit Sizes with PC/FAX Transmission

Unfortunately, PC/FAX boards don't respect the page makeup protocols in computer software. Page breaks aren't read, and the information is sent in a fashion that corresponds to the internal logic of the fax board, rather than to the way the message looks on the computer screen. Nor do most fax boards add their own page numbers.

This limitation creates several problems. First, some fax boards don't allow the sender to preview a message before it's sent. For others, the routine to do this is clumsy and awkward. Consequently, with PC/FAX, the sender often doesn't know exactly what the receiver has received, a shortcoming that can be exacerbated by the error correction schemes. For example, when some fax boards receive an error message after a page has been transmitted, the board will put the page aside, continue the transmission and only retransmit the faulty page at the end. If the pages aren't numbered, a great deal of confusion can result when the document is received.

Second, some fax boards automatically compress data generated in 132- and 200-column modes. Therefore, wide spreadsheets can be garbled.

FAX BOARD NETWORKING

Some fax boards are designed to work within networks, which can be very convenient for outgoing messages. Anyone on the network can create a document for fax transmission and send it to the board, which, in turn, will transmit the message to a fax machine anywhere in the world. Incoming messages, however, are less convenient. All incoming messages reside on one hard disk, and each recipient must check that disk periodically for messages. To receive and manage incoming fax messages adequately, the host computer must be equipped with a large hard disk and must be monitored regularly.

Some PC/FAX board vendors now offer products they claim will deliver a PC/FAX to a specific user on the network. Currently, this feature only works between boards made by the same manufacturer.

THE BIG QUESTION: DO YOU NEED PC/FAX?

Not surprisingly, your need for PC/FAX depends on the way fax is used within your organization. PC/FAX is most efficient as a sending terminal if people who wish to send fax messages generally create them on a personal computer.

For example, executives who write their own memos—or have them produced on a personal computer—may wish to have those memos faxed immediately to the receiver rather than wait for them to be printed and faxed by conventional means. The text image can be merged with a stored paint image of the company letterhead to produce a complete fax document.

Consultants, outside agencies and subcontractors, such as public relations and advertising agencies, who simply need customer approval for a job before they proceed, should explore PC/FAX if they create their roughs on a personal computer. Indeed, anyone who works primarily on a computer, generates heavy fax traffic and wants to send his or her own faxes should explore PC/FAX.

PC/FAX is also for you if you can exploit its power and flexibility. Remember, PC/FAX eliminates most limitations of stand-alone fax with respect to the number of fax numbers that can be stored in a directory, unattended operation, delayed transmission and so on. Although Chapter Seven warns you of the dangers of junk fax, PC/FAX can generate and send automatically hundreds of faxes to get your message across, and it can customize those faxes to look just like any other fax traffic.

Remember, however, that a fax message is appropriate only if you assume that the recipient will be satisfied with a hard copy of the information. If you wish the receiver to have data that can be manipulated, it's usually more efficient to establish a direct telecommunications link.

At the same time, fax can be a substitute, although not a very efficient one, in situations where telecommunications links are impractical or are too difficult to establish. For example, it's extremely difficult to communicate overseas electronically with a modem. With PC/FAX, it's extremely easy to communicate overseas at any time.

> **For your PC to serve efficiently as a receiving machine, it must run 24 hours a day.**

PC/FAX also can be efficiently installed in a local area network. With one fax board, all members of the network can send their own fax messages whenever they want to without relying on support staff.

On the other hand, PC/FAX doesn't serve equally well as a receiving machine, particularly when there's heavy fax traffic. Incoming messages on a PC not only require a lot of disk storage space, but may demand attention at an inconvenient time. In addition, the messages must be downloaded from your hard disk to your printer or into another storage medium. And unless the information is filed immediately, you risk losing it.

Finally, for your PC to serve efficiently as a receiving machine, it must run 24 hours a day. If you leave for a trip of any length, you must arrange for somebody to check your faxes regularly. Otherwise, you may not receive information your communications partners are sending.

Clearly, within a business environment, PC/FAX has pros and cons. It's an effective sending machine, but it requires a scanner in addition to the fax board to make it fully operational. As a sending machine, PC/FAX gives you additional control of your situation. As a receiving machine, it gives you less control because it demands too much attention at inappropriate times.

There's one situation, however, in which PC/FAX is invaluable—when you want to fax data which must be downloaded from a mainframe computer. With PC/FAX, the entire operation can be completely automated, requiring no user intervention. For example, in large retail chains, when an item is purchased, that information increasingly is sent from the cash register directly to a central computer which tracks inventory. Every night the central computer can download the performance and in-stock status of specific products to one or several personal computers that, in turn, can fax that information to the appropriate product manager or buyer. In distributing information that resides in large computers, PC/FAX offers tremendous efficiencies.

A Checklist: IS PC/FAX FOR YOU?

1. Do you intend mainly to send information via fax or do you plan to receive fax as well?

2. If you don't plan to receive information via PC/FAX, how will you receive incoming fax? Bear in mind that once you begin to send fax, you're likely to receive fax in return.

3. Is the information you plan to send generated in personal computers? If it isn't, will you have to buy a scanner, which adds to the cost of the system?

4. Can one computer station generate enough fax messages to justify the $1,000-plus investment? (It's as hard to share PC/FAX as it is to share a personal computer.)

5. Is the computer operator the right person to be a fax operator as well?

6. Is the information you want to send usable by the recipient in the form of hard copy? Would data be more useful?

7. Can you establish alternative electronics communications links with those with whom you wish to communicate?

8. Do you have much experience with stand-alone fax? Can you identify specific advantages PC/FAX will provide?

The answers to these questions should lead you to the right decision concerning PC/FAX. In many situations, PC/FAX is the perfect solution to distributing fax power throughout the office. Someday perhaps every computer will have a fax board.

7

A Look into the Future

Because fax technology will be an important element of our communication system for years to come, it's important to consider what the future might bring. Many "predictions" already exist, some of them through the combination of personal computers with fax. The real change will result when major fax manufacturers incorporate these enhancements into their main product lines.

RAPID BUT UNEVEN GROWTH

More than any other factor, growth will impact the future of fax. With fax machine sales rates projected to grow at 30 percent a year well into the 1990s, the sheer number of people with access to fax will shape its future most directly. As more people have fax, more people will need it. But at some point, perhaps within two or three years, growth will slacken off before the potential is fully realized. One buying group will be saturated, while another will have yet to enter the market.

> Soon you'll be able to buy a stripped-down Group III fax for $500.

> **Fax sales are projected to grow at 30 percent a year.**

Nevertheless, even with more than a million fax machines in place in the United States, this initial stage of growth should be dramatic and impressive and help make fax more universal.

COMPATIBILITY

Group III machines will last well into the 1990s. Although you'll hear many promises about Group 4 (e.g., 5-seconds-per-page transmission speeds and totally error-free transmission), the use of Group 4 machines requires a new type of network. Because telephone infrastructure won't change for many years, Group 3 fax is likely to serve you for a long time.

PRICING

Because so many companies are competing for a share of the market, fax buyers can look forward to continued pressure on manufacturers to provide more capabilities at the lowest price. Absolute prices may not drop far from current levels, but the features at each price point should increase.

For example, today you can buy a stripped-down Group 2 fax for around $500; soon you'll be able to buy a stripped-down Group 3 fax for the same price. At the same time, most buyers will probably opt for more feature-laden machines at a higher price.

The price categories outlined in Chapter Three also will change. Because many small businesses want machines with a lot of features, manufacturers will try to offer as many features as possible below those price levels.

But price decreases and the changed price/performance ratios won't be sufficient to warrant deferring purchase. Because fax isn't like the typical consumer-electronics item, prices aren't likely to plunge. Indeed, fax probably will follow the pattern set by personal computers. Although PC clones now cost as little as $800, most people still spend more than $2,000 for a system.

ERROR CORRECTION

In the short term, one major enhancement to fax machines will be the addition of Error Correction Mode (ECM). The CCITT has established a Group 3 optional ECM standard that allows different brands of machines to communicate with each other and detect errors during transmission. As each

> **Laser printing technology is likely to replace thermal printing.**

page is sent, the receiving machine is able to determine whether it has correctly received all the information. If it detects an error, it automatically requests the sending machine to resend the page.

INCREASED PAPER CAPACITY, LASER PRINTING AND BOND PAPER

The quality of paper and printing are weak areas in fax technology today, and many improvements are likely. Thermal printing is inexpensive, rugged and quick, and it lends itself to fax. But it doesn't have the look, feel or durability of regular bond paper.

As fax machine volume increases and as copier equipment continues to drop in price, laser printing technology is likely to replace thermal printing. Because inexpensive laser printers cost around $1,500 today, laser fax machines could sell for less than $2,000 in just a few years.

Ink-jet technology (such as that used in the Hewlett Packard DeskJet printer, which sells for less than $900) is also a viable fax printing technology. It uses plain paper to print, and its speed, though slower, is adequate for fax use.

In large companies, fax units with multiple bins may appear, allowing internal company memos to be printed on special paper or forms and documents from outside the company to be printed on another type of form.

INTEGRATION WITH OTHER OFFICE EQUIPMENT

When fax machines begin using laser printer technology, multipurpose machines will appear. Both computer laser printers and fax machines work efficiently in a group or departmental setting. A future combination fax/laser printer might have a port into which a document scanner can be plugged. The machine might be used in a local network to print computer documents or used as a sending and receiving fax.

Adding full-function copier capability is also a natural step. By adding document feeders and a scanner, these machines can be used for both fax and high-quality copier functions.

The internal memory used by fax machines also will be useful for copier functions. Original documents need be scanned only once; then the machine can print as many copies as needed from stored memory, resulting in less wear and tear on originals. Multiple output bins will be even more important as new multipurpose machines appear.

Although fax machines have telephone handsets, the trend isn't toward two line units, which would allow the handset to be the primary telephone as well.

Fax will continue to be a convenient and accessible technology, but not so accessible as to jeopardize security or double as a substitute telephone system.

INCREASED MEMORY

More memory will increase the flexibility of fax usage by allowing easier implementation of such advanced features as delayed transmission and broadcasting. Memory will be calculated in pages of information, and 50-page memories won't be uncommon. Memory also will be used to solve fax security problems.

> In the future, you'll be able to see the people you talk to on the screen.

When a document is received, only a cover sheet will be printed for the recipient. After the recipient enters a security code on the machine, the document will be printed from the fax machine's internal memory.

NEW COMPUTER-BASED MACHINES

Because computers will remain the center of office automation, PC/FAX will continue to proliferate. In addition to PCs with fax capability, a PC-based fax machine that performs only fax functions may appear, much like the Magnavox Video Writer, a PC-based dedicated word processor.

The trend in PC/FAX will be toward multifunction and multiport communication boards. These new boards will include fax features as well as computer modems, document-scanner ports and even ports for video images which will form the basis of communication boards.

Optical Character Recognition (OCR) will evolve into an integral part of PC/FAX. Once a fax is received, it will be converted to text data for storage or further processing. The ability to convert from image to data is a key element for the viability of PC/FAX. In the future, you'll be able to see the people you talk to on the screen and, at the same time, send faxes or other data on the same machine. The beginnings of this trend are now in development by some PC/FAX board manufacturers.

INTEGRATION WITH OTHER FORMS OF COMMUNICATION

As computer-to-computer electronic mail becomes increasingly important in large corporations, links to technology such as fax will become more common. Both AT&T Mail and MCI currently provide computer electronic-mail services to large companies.

For example, a buyer in a large company can fax requests for proposals to three vendors and send an E-mail copy of the same request to his or her other boss. The text needs to be entered only once, and the destination can be a computer or fax machine. Together, fax and E-mail will completely replace telex.

"FOURTH GENERATION" FAX

ISDN, which stands for Integrated Services Digital Network, will be adopted by the telephone system during the 1990s. But it has a long way to go before most of us will be affected by it.

Today's telephone network sends analog signals along phone lines. The new ISDN network will be designed for digital information, such as that used by computers and fax machines. In many ways, the difference between the two technologies is like that between phonograph records and compact disks (CDs). Records are analog, CDs are digital; and although CDs produce unmatched sound quality, they require new equipment.

Given the current size of the telephone network, the new equipment for ISDN will take many years to install. However, once ISDN is here, new Group 4 fax machines will be able to make use of its high-speed digital lines to speed error-free faxes in less than five seconds per page.

PORTABLE FAX

As more and more businesses rely on fax communication, portable fax machines will become popular with traveling executives. With a portable fax machine, travelers will be able to receive and review reports and memos in their hotel rooms. Field sales representatives will use portable fax together with cellular phones to send and receive information from their cars.

Portable fax can mean anything from small lightweight

machines requiring AC power to battery-operated units weighing just a few pounds. Like lap-top computers, portable fax machines will proliferate as people learn new ways to use them.

THE FUTURE OF FAX

Through professional education, thousands of new fax users will understand and then utilize fax in novel settings and situations. Companies will increase their fax applications, but the technology won't be restricted to the office environment. For example, as purchasing managers in various industries and professions rely on fax to obtain price quotes and to place orders, more and more suppliers will add fax capability to better serve their customers.

> Because it could be abused in a number of ways, fax isn't likely to become popular for home use.

Contractors may use fax to send documents to the building commission and get back comments or building permits without waiting for hours at city hall. Emergency prescriptions may be faxed to pharmacies, eliminating telephone calls and providing verification.

Because it could be abused in a number of ways, fax isn't likely to become popular for home use. Home fax units would be overrun with contest offers and solicitations. Because fax is less expensive than mail in a local telephone area, many companies would bill through fax if it were widespread. How would you feel if you paid for the fax paper each time you received a bill and didn't get a reply envelope for your check? The requirement for a dedicated phone line will also greatly limit home fax use.

At the office, although different departments will demand their own machines, faxes will not be found on every desk as telephones are today. PC users will use PC/FAX more and more if they're creating and sending large amounts of information. And personal faxes will be used by high-level executives (or their staff) who work with sensitive data and need to keep the information confidential.

As long as paper remains a convenient medium for sending and storing information, fax will continue to grow as an important communications technology. But it will grow in an evolutionary fashion. The essential process—the ability to scan, send and receive information on hard copy via telephone lines quickly and inexpensively—is already in place.

Although fax technology will improve, there's no reason to wait for the next generation of equipment if you have relevant fax applications now. In most cases, companies will add to their fax network incrementally, with international standards ensuring that current fax technology won't become obsolete. Although you can expect fax technology to be improved five years from now, the current state of technology is good enough to have a dramatic and beneficial impact on your communications infrastructure.

TO LEASE OR TO BUY

As with other types of office equipment, many companies opt to lease fax machines rather than buy them. Most major suppliers provide leases for their larger machines, and office equipment dealers generally offer leases through third-party leasing companies.

There are two major types of leases—a financial lease and operating lease. A financial lease resembles traditional borrowing. The lessee makes a series of payments which at least equal the full price of the machine. The lease cannot be canceled. At the end of the lease period, the lessee owns the machine.

Financial leases are popular with small businesses for several reasons. Because lessors are, in essence, lending money against a specific piece of equipment (the fax machine), and because the lessors own the machine until the lease ends, lessors are better protected against nonpayment than general lending institutions. Consequently, it may be easier for a small business to lease equipment than to borrow money to buy the machine.

A business usually can get a lease for 100 percent of the selling price of the equipment. However, the rates on a financial lease may be higher than the rates offered by other lending sources.

The second type of lease, an operating lease, most resembles equipment rental. An operating lease protects a company against the obsolescence of equipment, because operating lease payments typically don't cover the full price of the machine. At the end of the lease period, however, the lessor still owns the equipment and can lease or sell it to another party, while the lessee can choose a more advanced model. While most operating leases can be canceled, the cancellation terms can be so onerous that buying the machine outright in some situations is more practical.

Leasing also has tax implications. For example, on leased machines, companies can take a research and development tax credit which otherwise isn't allowed. Moreover, companies subject to the alternative minimum tax can accrue certain tax advantages through leasing. Because every company's situation is different, consult with your accountant to determine your course of action.

In evaluating leasing options, remember that a stripped-down Group 3 fax retails for less than $1,000, and a top-of-the-line unit with all the bells and whistles may cost only $5,000. As a result, it can be more efficient just to borrow the money to buy a machine or a few machines outright.

PUBLIC FAX CENTERS

Despite the explosive growth of fax, many companies still don't have fax machines. Indeed, you may have decided that you don't wish to invest in fax technology right now. Your volume of fax traffic may be too low, or you may want to wait for the next generation of machines.

To meet the needs of those who only occasionally use fax, many business service centers, mailbox kiosks and quick copy centers have begun to offer fax facilities to the public. Though expensive, the services are easy to use. To send, you simply bring your document to the center and an operator does the rest. To receive, you simply tell the sending party the telephone number of the fax center you plan to use. Be sure the sender includes your name and telephone number on the transmittal sheet.

As demand increases, the price for fax services will fall dramatically. The services generally use one of two billing methods. The first is a per-page charge that includes the telephone charges that the service incurs. Some typical rates are as follows:

Incoming faxes — $1.00 per page.

Outgoing faxes —

a) within the U.S. and Canada — $2.00 per page for the first three pages, $3.00 for each subsequent page (the transmittal sheet counts as a page, and pages refer to standard 8.5 x 11 inch letter-size paper; 8.5 x 14 inch legal-size pages cost 50 percent more).

b) international — $10.00 for the first page, $7.00 for the second and $5.00 for each additional page.

The second method of billing used by fax services is a per-page rate for the service and a separate bill for the associated telephone charges. With this method, typical rates for sending or receiving are $2.00 for the first page and $1.00 for each additional page, plus phone charges.

For someone whose fax use is limited and infrequent, using a fax service provides an alternative to owning a fax machine. It also allows you to fax materials to people who don't own a machine. Clearly, for documents of five pages or less, even fax services represent a competitive alternative to overnight mail. On the other hand, if you routinely use a fax service, you may want to investigate buying your own fax equipment.

Index

fax announcement, 86
fax boards, 105-107
fax directory, 26-27, 77, 94-95, 99
fax growth, 119
fax junk mail, 94-96
fax manager, 80, 100
fax number, 26, 77, 86, 95, 99
fine resolution, 18, 22
foreground, 109

graphics, 7, 9, 104-108
Group 1, 14-16, 47, 48, 62
Group 2, 14-16, 47, 48, 61, 62, 105, 120
Group 3, 14-17, 21, 41, 46, 47, 48, 52, 61, 62, 105, 107, 108, 120
Group 4, 14-17, 63, 64, 120

halftone transmission, 32, 41, 51, 52
handshaking, 53

inappropriate faxes, 90-91
information management, 84
installation, 78-80
internal memory, 34, 37, 38, 50, 54
invoice, 11, 12, 43
ISDN, 14, 17, 63, 124

junk fax, 95-96

laser printer, 18-19, 59, 100, 108, 121, 122
LCD display, 27
lease versus buy, 127
legal, 97

machine location, 72-75
mail, 3-4
maintenance, 20, 100
management features, 23, 42-43
manual receive, 24
manufacturers' sales reps, 64, 78
messengers, 5-6
modem, 21, 22, 32, 106

networking, 113

OCR, 106-107
office equipment stores, 65, 79
one-touch dial, 25, 54
operator, 21, 80, 92, 98
optical mark reader (OMR) sheets, 39
overnight mail, 5-6, 56, 60, 95, 96, 130

Want to Know More?

For a list of major fax machine vendors, fax-related periodicals, research groups and organizations, send $1.25 (check only) to:
Ventana Press, P.O. Box 2468, Chapel Hill, NC 27515.

___Please send me the list of fax vendors and organizations. I've enclosed $1.25 (no cash please).

Name_____

Firm_____

Address_____

City_____State_____Zip_____

Telephone_____

___Please add my name to your mailing list for future titles.

Enhance your productivity with two books from the Ventana Press Desktop Design Series

$23.95, plus
$2.40 shipping
241 pages,
220 illustrations

Looking Good in Print: A Guide to Basic Design for Desktop Publishing
by Roger C. Parker
No matter which software you use, this book will help you become a better designer. Hundreds of tips and tricks to make your printed materials look better.

$21.95, plus
$2.40 shipping
366 pages,
250 illustrations

Desktop Publishing with WordPerfect, also by Roger C. Parker
You can make any document more appealing by mastering WordPerfect 5.0. Its graphic capabilities will save you time and money over traditional design and paste-up methods.

Your money returned if not satisfied. Available at bookstores or from
Ventana Press
P.O. Box 2468
Chapel Hill, NC 27515
919/942-0220

LIBRARY
ST. LOUIS COMMUNITY COLLEGE
AT FLORISSANT VALLEY.

Please send me ___ copy(ies) of *The Book of Fax* at $12.95 per book. Add $2.40 a book for regular UPS shipping; $5 for two-day air. NC residents add 5% sales tax. Immediate shipment guaranteed.

Name_____Firm_____
Address (no P.O. Box)_____
City_____State_____Zip_____
Telephone_____
___Check or money order enclosed.
___Visa or ___MC
Acct.#_____
Exp. Date_____
Signature_____
Ventana Press, P.O.Box 2468, Chapel Hill, NC 27515
919-942-0220

Please send me ___ copy(ies) of *The Book of Fax* at $12.95 per book. Add $2.40 a book for regular UPS shipping; $5 for two-day air. NC residents add 5% sales tax. Immediate shipment guaranteed.

Name_____Firm_____
Address (no P.O. Box)_____
City_____State_____Zip_____
Telephone_____
___Check or money order enclosed.
___Visa or ___MC
Acct.#_____
Exp. Date_____
Signature_____
Ventana Press, P.O.Box 2468, Chapel Hill, NC 27515
919-942-0220

NO POSTAGE
NECESSARY
IF MAILED
IN THE
UNITED STATES

BUSINESS REPLY MAIL

FIRST CLASS MAIL   PERMIT # 495   CHAPEL HILL, NC

POSTAGE WILL BE PAID BY ADDRESSEE

Ventana Press

P.O. Box 2468

Chapel Hill, NC 27515

NO POSTAGE
NECESSARY
IF MAILED
IN THE
UNITED STATES

BUSINESS REPLY MAIL

FIRST CLASS MAIL   PERMIT # 495   CHAPEL HILL, NC

POSTAGE WILL BE PAID BY ADDRESSEE

Ventana Press

P.O. Box 2468

Chapel Hill, NC 27515